GLASSES AND THEIR APPLICATIONS

GLASSES AND THEIR APPLICATIONS

H. RAWSON

The Institute of Metals

1991

Book Number 499

Published in 1991 by The Institute of Metals
1 Carlton House Terrace, London SW1Y 5DB

and
The Institute of Metals
North American Publications Center
Old Post Road, Brookfield VT 05036
USA

British Library Cataloging in Publication Data

Rawson, Harold
    Glasses and their Applications.
    I. Title
    666.1

IBSN 0 901462 89 6

Library of Congress Cataloguing in Publication Data
Applied for

Text design: PicA Publishing Services, Abingdon

Technical illustrations: Paul Burnell

Cover design: Jeni Liddle

Typeset by LaserSet, Abingdon, Oxon

Made and printed in Great Britain by Bell & Bain Limited, Glasgow

Contents

viii

Foreword

In the preface to an earlier book, published in 1980, I remarked that there was no need to apologise for writing another book about glass. There were so few of them that one more was certain to be welcome. That may have been fair comment at the time, but it is certainly not true now. There are a number of excellent books and many valuable review articles, primarily concerned with various aspects of glass science or with new materials of current research interest. There are also several books dealing with the manufacturing processes used in the industry.

Discussions between the Publications Committee of the Institute of Metals and myself led to the conclusion that it would be of value to produce a book giving a wide ranging survey of the applications of glasses which would include both the well established uses in containers and in glazing where the annual requirement runs into millions of tonnes and applications which are still under development for which the annual requirement is never likely to be more than a few tonnes. Even so that small quantity of material might play a very important role, as in fibre optic communications for example.

Many interesting developments are taking place involving applications of glasses. Some involve new materials, such as metallic glasses and sol–gel materials. Some have arisen from technical developments in other fields – the use of glass for thee disposal of nuclear wastes being one example. Less revolutionary, but very important commercially and in other ways, is the development of low-emissivity and solar control coatings for float glass. The only part of the glass industry for which the future is uncertain is the glass container industry. However several large international R. & D. programmes have been under way in recent years with the aim of ensuring that the industry loses no more of its present market share to competing container materials.

So this is a book concerned with technology, but not with manufacturing technology, except when a brief explanation of manufacturing processes is needed before one can begin to explain the technical background of the application.

To compress the large amount of information available within the limits of the Institute's word count has been difficult, and eventually impossible. The intention has been that the broad coverage should make the book useful to a reader with a general interest in materials and their applications, but the many references especially to review articles and specialised texts would direct the reader with some specific interest to sources giving a comprehensive treatment of any particular topic. The result is no more than a sketch map, but with many signposts.

When surveying the literature on so many different applications, it is difficult to strike a sensible balance. In this field, as in others, a number of new developments have taken place with prospects of new and commercially very profitable applications – even though these may not yet have materialised. The literature in these fields is very extensive. However when one turns to the well established applications in containers and in glazing, there is relatively little recent literature. It would not be very sensible however to rate the importance of any activity by the extent of the associated literature or, at the other extreme, to allocate so many pages per Megatonne of glass melted.

In any field of the science and technology of materials, one needs some understanding of the nature and structure of the material and of the factors which determine its properties. This understanding does not need to be very deep or very detailed if one's interest is technological. I hope Chapters I and II will cover enough of that background for most readers. Because mechanical properties are very important in many applications, an entire chapter is devoted to this subject, again with a technological emphasis.

I occasionally refer to national and international standards for methods of test or product performance. Usually I give only the British Standard number and title. It is easy enough in the UK to identify corresponding foreign standards when necessary and I assume that readers in other countries will be able to do the same.

In the preface which I referred to at the beginning, I apologised to my children for spending time writing a book instead of caring for their interests. I am less inclined now to apologise to my grandson, who at present is more likely to regard a book as a missile rather than as a medium of instruction or entertainment. I hope that each reader of this book will find at least one paragraph or one reference of enough value to restrain him or her from joining the missile throwing faction and that someone will be so dissatisfied as to immediately set about writing a far better book.

Sheffield
January 1990

Acknowledgements

I am grateful to the following for permission to use figures and tables:

The Society of Glass Technology: Figures 4, 5, 16, 23a, 24, 27, 34, 38, 63, 64, 67, 68, 96, 99, 102, 108, 114. Tables 13, 14, 15, 16.

The American Ceramic Society: Figures 3, 7, 12, 14, 17, 18, 26, 31, 32, 36, 39, 43, 76, 139. Table 3

American Journal of Science: Figure 22.

Deutsche Glastechnische Gesellschaft: Figures 21, 46, 80, 81, 92.

Friedrich–Schiller University: Figures 107, 112.

American Chemical Society: Figure 2.

Institute of Electrical Engineers: Figures 95, 113.

American Institute of Physics: Figures 11, 41, 42, 100, 111.

American Society of Testing and Materials: Figures 49, 50.

Optical Society of America: Figure 104.

The Electrochemical Society: Figure 109.

Institute of Electrical and Electronic Engineers: Figures 115, 116.

Union Scientifique Continentale du Verre: Figure 45.

Corning Incorporated: Figures 10, 15, 29, 61.

General Electric Co.: Figure 98.

Pilkington Glass Ltd.: Figures 56, 57, 58. Table 11.

Schott Glaswerke: Figures 19, 69, 72, 73, 74.

Academic Press: Figures 30, 137, 138. Table 18.

Ashlee Publishing Co.: Figures 20, 35, 53, 77, 78, 20, 51a, 51b. Table 4.

British Standards Institution: Figures 52, 117.

Chapman and Hall: Figure 47, 101.

Elsevier Science Publishers: Figures 6, 9, 28, 48, 82, 103, 55, 75, 79, 86, 87, 88, 89, 90, 93, 110, 121, 122, 123, 125, 126, 127, 129. Tables 12, 18.

Hutchinson Benham Ltd.: Figures 59, 60.

Illiffe Press: Figure 23b.

Martinus Nijhoff Publishers: Figures 37, 44, 119, 120, 124, 130, 135, 136. Tables 17, 19, 22, 23, 27.

Springer–Verlag: Figures 8, 13.

Pergamon Press: Figures 33, 118.

Pitman Ltd.: Figure 97.

Plenum Press: Figure 62.

Taylor and Francis Ltd.: Tables 6, 20, 21.

Van Nostrand–Reinhold: Figures 131, 132, 133. Tables 24, 25, 26.

Wiley–Interscience: Figures 40, 128.

Mrs D. F. Horne: Figure 106.

Chapter 1

The Nature of Glass

1.1. Introduction

Even in a predominantly technological book like this one, it is necessary to give a brief account of the nature of glass and to review those aspects of the basic characteristics of the material which help one to understand its behaviour and properties. A number of texts and symposia proceedings are available which cover various aspects of glass science in considerable detail (Wong and Angell 1976; Elliott 1983; Zallen 1983; Uhlmann and Kreidl 1983; Wright and Dupuy 1985; Scholze 1988).

Almost all glasses of commercial importance contain more than 60wt-% of silica. The raw materials, after mixing, are fed into a furnace which heats the mixture to a temperature of 1500°C or more. Even at the maximum temperature in the furnace, the melt is very viscous (ca 10 Pa s compared with 1.5 Pa s for glycerol at room temperature). Because of the high viscosity, it takes many hours for bubbles to disappear from the melt and for it to become sufficiently homogeneous. The melt finally passes to a section of the furnace where its temperature is reduced to a uniform lower value (typically 1100°C). Its viscosity is then about 1000 Pa s, a value suitable for the start of the shaping operation, of which there are many different kinds. As the melt is shaped, it is cooled further until the viscosity of the surface layers reaches a value of 10^{11} Pa s or more, when the glassware is sufficiently rigid not to deform under its own weight. Finally from about 600°C down to room temperature, the glassware is subjected to a slow cooling process called annealing. This allows any stresses present in the ware to decay by viscous flow and also ensures that the glass is not subjected to temperature gradients which will cause permanent stresses in the glass when it reaches room temperature. (The viscosity values given are typical for a container or flat glass composition. Chapter 2 discusses in a more general way the effects on viscosity of glass composition and temperature.)

Note that in the above account the word 'melt' is used when referring to the material at high temperatures in the furnace and the word 'glass' appears only when referring to the material at low temperatures. The distinction may seem pedantic, but it expresses a real difference as will be seen later.

This brief account was written with the mass production processes of the industry in mind. It has little obvious relevance to the manufacture of less conventional glasses and products made from them. Thus metallic glasses are made on a small scale from very fluid melts. Cooling is extremely rapid and no annealing of the conventional kind is required. In making high silica fibres for optical communication applications, a major part of the glass fibre is formed by a vapour phase deposition process. No furnace is involved – at least not one which acts as a container for the material. Also there are oxide glasses of some practical importance which are made from quite fluid melts, the viscosities of which increase so rapidly with decreasing temperature that they cannot be shaped by any of the processes commonly used in the industry.

This wide range of materials and processes makes it difficult to give a simple definition of the term 'glass'. The conventional way of dealing with this problem is to give an account of what is true of the vast majority of glass-forming materials and consider later materials that appear to be exceptions.

1.2. The melt – glass transition

The volume-temperature diagram (Fig. 1) is frequently used when discussing the relationships between a melt, a supercooled liquid and a glass. Most liquids crystallise rapidly at a well-defined temperature, T_f (the melting point or liquidus temperature) with a marked change in volume – usually a decrease. If the melt is completely free from crystal nuclei or foreign particles, it can be supercooled to some extent, the volume following the line b–e. Glass-forming melts can be supercooled to an unusually high degree, even when nuclei are present. Usually, but not always, these are melts which have a relatively high viscosity at the liquidus temperature. The supercooled melt is thermodynamically unstable. Crystallisation will result in a decrease in free energy and, given time, this is what will happen.

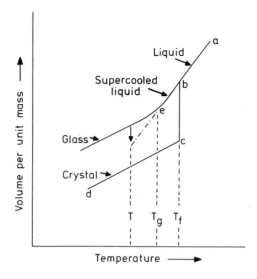

1. Relation between the glassy, liquid and solid states

The viscosity of the supercooled melt continues to increase as the temperature is reduced until a temperature is reached, near T_g on the diagram, below which the material is for most practical purposes a solid. This temperature range around T_g is called the transformation range and the viscosity is then about 10^{12} Pa s. Only below T_g is it correct to refer to the material as a glass. The more or less marked change in slope in the volume–temperature curve indicates that in this temperature region the structure of the melt becomes relatively insensitive to further reductions in temperature. At room temperature, the glass has a structure similar to that which was frozen in at T_g. Below T_g a glass, like a supercooled liquid, has a higher free energy than a crystalline phase or a mixture of crystalline phases. However, because structural rearrangements can occur only extremely slowly at temperatures well below T_g, the glass is stable for all practical purposes.

The room temperature properties of a glass are affected by the rate of cooling through the transformation range. The figure indicates that if the supercooled melt is cooled slowly, the volume – temperature curve follows the dotted line below e. Consequently the glass at room temperature is slightly more dense than if it were cooled rapidly. This is very important in some applications of glass. Thus optical glasses must be cooled at an extremely well controlled rate, otherwise the refractive index will vary from one part of a block to another due to differences in thermal history. A conventional annealing schedule, which is required to do no more than keep internal stresses at a low level, is not good enough to ensure adequate control of the refractive index.

Heat treatment above the transformation range, but below the liquidus temperature, may also have a marked effect on material properties, as will be seen later.

The liquidus and transformation temperatures vary considerably with composition. At this stage it may be helpful to have one or two values in mind. For commercial soda-lime-silica glasses, such as container glass or float glass, the liquidus temperature is about 950°C and the transformation temperature 500°C. For pure silica the melting point of the crystalline phase is 1720°C and the transformation temperature of the glass is about 1100°C.

1.3. Glass-forming substances

Glasses can be made from many substances, both organic and inorganic. This section will concentrate on inorganic glass-forming systems and especially on those which are (or which may become) technologically important.

Glass artefacts were first made in ancient Egypt and, since the Iron Age, in Europe (Renfrew 1976). All were made from naturally occurring raw materials and contain a major percentage of silica, which is the glass-forming oxide, in the sense that silica is the essential component for glass formation to occur. It was not until the nineteenth century that there was any systematic attempt to find and to study new glass-forming systems. This led, in particular, to the discovery of phosphate and borate glasses and the recognition that boric oxide and phosphorus pentoxide are themselves glass-forming oxides. Later germanium dioxide and arsenic trioxide were added to the list.

That list would probably have been regarded as complete as recently as 1950 and, even now, most commercial glass compositions are based on silica, boric oxide, alumina and a number of basic oxides – alkali and alkaline earth oxides.

Since then, there has been a considerable increase in our knowledge of glass-forming systems. Table 1 lists the more important glass-forming systems which have been discovered and studied during the past thirty years.

An interesting feature of the new oxide glasses listed in the table, i.e. the binary aluminate, tellurite and vanadate glasses, is that neither component when melted alone will readily form a glass, yet when the components are melted together in suitable proportions, glasses are easily formed.

Turning to the non-oxide glasses: it has been known for many years that the element selenium readily forms a glass, as does arsenic selenide and arsenic sulphide. Arsenic sulphide glass became commercially available in the 1960s as an infrared transmitting window material. Then, mainly in the Soviet Union, a considerable amount of work was carried out on glasses based on the elements S, Se and Te, which are now known as chalcogenide glasses. Interest in these systems developed rapidly, partly because many have excellent infra-red transmission out to beyond a wavelength of 10 μm. Other chalcogenide glasses are semiconductors, which have been intensively studied partly because they can be made into devices with high speed switching characteristics.

In 1960 metallic glasses were discovered (Duwez *et al.* 1960) but this had little impact at the time, possibly because it was difficult to imagine how metallic glasses could be made in any usable quantity in view of the very high cooling rates required. Since then, they have been intensively studied and applications are being actively developed e.g. as magnetic and as high-strength materials. This was an unexpected development, as was the current intensive research activity on halide glasses.

Halide glasses based on beryllium fluoride and zinc fluoride have been known for many years but in 1974 Poulain accidentally discovered a wide range of glasses based on the so-called heavy metal fluorides, ThF_4 and ZrF_4 and LaF_3 (Poulain *et al.* 1975). These transmit to longer wavelengths in the infrared than do the silicate glasses and may eventually replace silica-based fibres in optical communication systems.

The present author reviewed the current state of knowledge of inorganic glass-forming systems in 1967. A more extensive review has been published by Kreidl (1983); van der Sande and Freed (1983) have reviewed metallic glass systems, and reviews by Tran *et al.* (1984) and by Drexhage (1985) deal with the heavy metal fluoride systems.

Table 1
Examples of newly discovered glass-forming systems

A. Oxide systems
    Al_2O_3–CaO
    TeO_2–BaO, PbO
    V_2O_5–BaO, PbO
    Tungstate, titanate and molybdate systems

B. Halide systems
    $ZnCl_2$
    $ZnBr_2$
    BeF_2 – RF, RF_2
    Heavy metal fluorides – based on ZrF_4, LaF_3, and ThF_4

C. Chalcogenide glasses
    Se
    As-S
    As-Se
    Si–Te
    Ge–Se
    As-Te-I
    Ge-Te-I

D. Ionic salts and aqueous solutions
    KNO_3–$Ca(NO_3)_2$
    KAc–$CaAc_2$ where Ac=acetate
    K_2SO_4–$ZnSO_4$
    $Ca(NO_3)_2 \cdot 4H_2O$
    Nitrate and chloride solutions

E. Metallic systems
    Au–Si
    Pd–Si
    Ni–P
    Fe–B

1.4. The structure of glasses

As stated earlier, the structure of a supercooled liquid is frozen in as it is cooled through the transformation range. A glass is therefore an amorphous solid with a structure similar to that of the same material at T_g.

It is considerably more difficult to determine the structure of a glass or even that of a 'simple' liquid like water, than that of a crystalline material. Much work has been done on the structure of glasses during the past thirty years, e.g. by X-ray and neutron diffraction and by a number of spectroscopic methods. Computer modelling methods, especially the molecular dynamics technique, have been developed which, considering the lack of some basic information about the properties of the interatomic bonds, give structural information which agrees surprisingly well with that obtained experimentally (Soules 1989).

Current knowledge of glass structure is a partial synthesis of the information obtained by many methods. Structural features which involve only nearest atom neighbours can be determined with reasonable accuracy. It is more difficult to describe the medium and long range structural features around any specified reference atom (Wright 1989; Elliott 1989).

Fortunately many of the properties of the technologically important oxide glasses can be discussed in terms of a simple model introduced by Zachariasen in 1932 and subsequently confirmed by Warren and his co-workers using X-ray diffraction. Some additional features have to be incorporated , based on more recent studies. These involve changes in the co-ordination numbers of some network-forming elements and are discussed later.

Zachariasen's ideas about the structures of glass-forming oxides were strongly influenced by the then current knowledge of the structures of the crystalline forms of these materials and by a general understanding of the principles of crystal chemistry. The most important glass-forming oxides are characterised by relatively small cations, Si^{4+}, B^{3+} and P^{5+}, each surrounded by either three or four oxygen ions. The basic unit of structure is therefore a triangle or tetrahedron. In the pure oxide, the units are joined together at their corners only and each oxygen ion is shared by two cations. This arrangement is a common feature of the corresponding crystalline materials, as was well known from the work of X-ray crystallographers such as Zachariasen himself and the Braggs. Thus Zachariasen simply proposed that a particular glass would have a structure built from the same or similar units to those found in the corresponding crystalline compound. However whereas in the crystal, unit cells are regularly repeated throughout the material, in glasses there is a variability of bond angles and of bond lengths. His proposal that glass structures are governed by the principles of crystal chemistry, though a simple one, had a revolutionary effect on glass science. (This can readily be appreciated by reading papers in this field published in the 1920s.) It has dominated thinking about glass structure ever since.

Zachariasen's requirements for small cations, low co-ordination numbers and corner sharing of units lead to a relatively open network structure, as shown in Fig. 2 which compares the structures of the same material in its crystalline and vitreous forms. He believed that such an open structure was necessary in order to accommodate the disorder associated with the glassy state without significantly increasing the lattice energy. He also believed that glass formation would be possible only if the disordered form had a lattice energy not greatly in excess of that of the corresponding crystal.

Zachariasen summarised his ideas in the form of a number of rules which must be obeyed if an oxide A_mO_n is to be a glass former:

1. No oxygen atom may be linked to more than two A atoms.
2. The number of oxygen atoms surrounding A atoms must be small.
3. The oxygen polyhedra must share corners with each other, not edges or faces.
4. At least three corners of each polyhedron must be shared.

The success of the Zachariasen model had one possibly negative effect. His 'explanation' of why open network structures should be necessary for glass formation was too readily accepted. Even when his paper was published, glasses had been described in the literature (but overlooked), e.g. aluminate, vanadate and tellurite glasses which do not readily fit into the rules which he proposed. It was not until much later that a more open-minded search for new oxide glass-forming systems began.

The subsequent experimental work of Warren included both two- and three-component glasses, in which basic oxides containing relatively large cations were introduced. A simple example is the system Na_2O–SiO_2. Knowledge of the structure of crystalline silicates was again a useful guide as to what type of structure to expect. The additional oxygen ions are accommodated by the breaking of –Si–O–Si–bridges which link the tetrahedra in silica glass. For each additional oxygen introduced, two 'non-bridging' oxygen ions are formed (Fig. 3). The sodium ions are accommodated in the relatively large holes in the network, surrounding themselves with, on the average, six

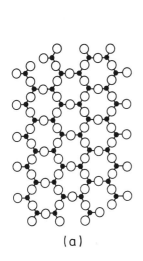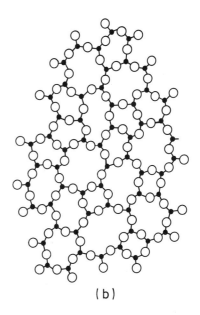

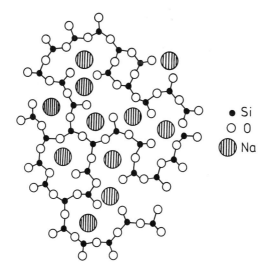

2. Structure of the compound A_2O_3 in (a) the crystalline and (b) the glassy form
(Zachariasen 1932)

• Si
○ O
Na

3. Structure of a Na_2O–SiO_2 glass

oxygens; a number to be expected from their size relative to that of the oxygen ion. The negative charge of the additional oxygen ion is localised mainly on the two 'non-bridging' oxygen ions that have been formed. In this way the positive charge of the sodium ion is locally neutralized.

It should not be assumed that the basic oxide is incorporated into the network with no, or only a local, disturbance or that the basic cations are distributed completely uniformly throughout the structure. The modifying cations will satisfy their own local structural requirements and, in some systems, may show a tendency to cluster.

As in the crystalline oxides, there are some cations which can exist in more than one co-ordination state e.g. Al^{3+} (six or four). B^{3+} (three or four). If these cations are present in the glass, they change their co-ordination number as a result of changes in the composition of the glass. These structural changes are reflected in the ways in which glass properties change with composition (Chapter 2).

Fig. 4, based on NMR results, shows that up to about 30 mol % Na_2O in the $Na_2O–B_2O_3$ system the addition of alkali converts boron atoms from 3 to 4 co-ordination. Within this region of low alkali content, a number of techniques have shown that the borate network is characterised by well defined structural groups each containing a particular number of BO_3 triangles and BO_4 tetrahedra. The relative proportions of these groups change as the alkali content increases (Griscom 1978a).

When the percentage of basic oxide, e.g. Na_2O, is progressively increased in the silicate systems, the number of non-bridging oxygens increases correspondingly and the continuous network is broken down. At the ortho-silicate composition, when the O/Si ratio is four, one would expect to find only isolated tetrahedra and so glass-formation would be unlikely. In fact it is very difficult in the sodium silicate system to make glasses containing more Na_2O than the metasilicate composition, for which the Si/O ratio is only 3. Glasses can however be made from compositions involving a mixture of alkalis with a O/Si ratio much higher than that of the metasilicate (Trap and Stevels 1959).

The breakdown of the network which occurs when alkali oxides are added to a silicate glass markedly affects its properties. In particular, the viscosity at any specified temperature decreases and the thermal expansion coefficient increases. The modifying cations play an important role in determining other properties. The ions are held by relatively weak ionic bonds to the network and hence are more mobile than the network-forming cations. They determine the ionic conductivity of glasses and play an important role in reactions involving diffusion, e.g. attack by aqueous solutions and other reactions involving ion-exchange.

The variation of glass properties with composition in alkali borate systems is quite different, as will be seen in the next chapter.

Table 2
Compositions of some commercial glasses (weight per cent)
(Major components only)

	SiO_2	Al_2O_3	B_2O_3	MgO	CaO	PbO	Na_2O	K_2O		
1	71.8	1.01		3.76	8.78		14.6	0.6		
2	72.0	1.3		3.5	8.2		14.3			
3	71.5	2.0		2.8	6.6		15.5	1.0		
4	56.0					29.0	2.0	13.0		
5	80.8	2.2	12.0	0.3	0.3		4.2	0.6		
6	67.5	2.5	21.7				3.2	4.2		
7	75.5	2.6	16.0				3.7	1.7		
8	52.9	14.5	9.2	4.4	17.4		1.0			
9	67.0	5.0					7.0	8.3	BaO	11.7
									Li_2O	0.6
10	68.3	0.2	2.2	4.6		2.9	14.4	7.0		
11	26.9	0.5				71.3	1.0			
12			20.0						La_2O_3	36.0
									Ta_2O_5	28.0
									ThO_2	16.0
13	5.5	17.5	16.0		9.5				BaO	52.0
14	5.0		17.0			64.0			ZnO	14.0

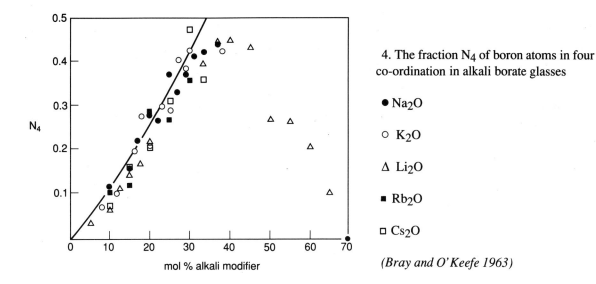

4. The fraction N_4 of boron atoms in four co-ordination in alkali borate glasses

● Na_2O

○ K_2O

△ Li_2O

■ Rb_2O

□ Cs_2O

(Bray and O'Keefe 1963)

1.5. Some commercial glasses and their properties

Table 2 gives the compositions of a number of important commercial glasses. The intention is merely to indicate the general features of each. For any particular type, glass compositions vary somewhat from one manufacturer to another. Note that most glasses contain many oxide components. This is largely due to the fact that any commercial glass has to meet a number of requirements, some imposed by its applications and others by the manufacturing process.

The first group of three glasses (1 to 3) are respectively float glass, container glass, and the glass used to make incandescent lamp bulbs. They are very similar in composition. They have suitable viscosity-temperature characteristics for the fabrication (shaping) processes which are used to make the glassware required. They have adequate chemical durability and the raw material costs are relatively low. Apart from meeting these main requirements, there are no other important property requirements to be met, except that for glass 3 the thermal expansion coefficient has to be carefully controlled.

Glass 4 is the traditional lead crystal composition, in which the high PbO content gives a high refractive index and a brilliant appearance to the glass when it is cut. A similar composition is used to make the high electrical resistivity glass for the 'stem' in incandescent lamps.

Glasses 5 to 7 represent the large and important group of borosilicate glasses. Composition 5 is widely used to make chemical apparatus and domestic ovenware. It has good resistance to thermal shock on account of its low thermal expansion coefficient and also has excellent resistance to chemical attack, especially by acids. Compositions 6 and 7 are sealing glasses with expansion coefficients matched to those of particular metals. Glass 6 seals to Kovar, Nilo K or Fernico, and is widely used both in large seals of tubular design and also in small terminal seals. Glass 7 has a somewhat lower expansion coefficient and seals to tungsten.

Glass 8 is one of a group of high melting point, low alkali, aluminosilicate glasses. It is commonly known as E glass and is made as continuous fibre for glass fibre reinforced composite materials and for weaving into glass textiles used in electrical insulation.

Glass 9 is a TV tube glass composition.

Glasses 10 and 11 are optical glasses, 10 having a refractive index of 1.518 and 11 of 1.805. The unusual composition 12 is also an optical glass (refractive index 1.85), the use of which makes it possible to design more highly corrected lens systems than is possible with the high lead glasses.

Glass 13 is resistant to attack by sodium vapour at elevated temperatures and is used as a coating on the inner wall of the glass arc tube in low pressure sodium vapour street lamps.

Glass 14 is a 'solder' glass. It is quite fluid at about 550°C and is used to seal together, without distortion, components made from soda-lime glasses such as composition 3. It is necessary, of course, that the expansion coefficient of the solder glass should match that of the components to be joined together.

Further information on commercial glass compositions is given by Volf (1961); Tooley (1984); Bansal and Doremus (1986) and Boyd and MacDowell (1986).

1.6. Devitrification

We have seen that a glass at temperatures below T_g is in a state of metastable equilibrium. If it is heated to a temperature between T_g and the liquidus temperature, it will tend to change to a condition of lower free energy. Eventually it will devitrify, i.e. convert to a crystalline phase or a mixture of such phases. Devitrification normally occurs from the glass surface where many nuclei are available. In time, the surface becomes covered with an opaque layer and the material is then of no value.

For most commercial glass compositions, devitrification is rarely a problem, even when carrying out processes such as hot bending (in the manufacture of automobile windscreens) and in the sealing operations used in the lamp industry. Devitrification problems are however occasionally encountered in the glass melting process itself, especially in that part of the furnace where the temperature of the glass is being brought down to the level at which it is fed to the machine. The flat glass industry was at one time more troubled by this than the container industry and for that reason flat glass compositions were chosen to be slightly different from container compositions in order to reduce the problem.

Fig. 5 shows the results of an inter-laboratory exercise to determine the effect of temperature on the growth rate of several crystalline phases in a sheet glass composition. All glasses show this shape of curve, differing only in the values of the growth rates and in the temperature range involved. The rapid rise in growth rate below the liquidus temperature is an indication of the increasing thermodynamic driving force, i.e. as the temperature falls the free energy of the more stable mixture of crystalline phases becomes increasingly less than that of the glass. As the temperature falls further, the rate of crystal growth decreases due to kinetic factors i.e. the structural changes occur more slowly because of the decreasing diffusivities of the ions involved, resulting from the decreasing thermal energy available to make the necessary structural rearrangements.

The understanding of the factors which control the rates of nucleation and crystal growth is fundamental to the understanding of glass formation itself. Fig. 6 shows schematically the way in which the rate of crystal growth and the rate of nucleation vary with the temperature. Obviously, for glass formation to occur, the maximum values of the rate of nucleation and the rate of nucleation should be low. Maximum crystal growth rates range from 10^{-7} m/s for a commercial flat glass composition to 1 m/s for a metallic glass. Clearly the metallic glasses require very rapid cooling of the melt and can only be made in the form of very thin ribbon or sheet.

The following equations describe the temperature variation of the nucleation rate and the rate of crystal growth. They contain both thermodynamic and kinetic terms.

The rate of nucleation, I, is given by:

$$I = A.\exp(-W^*/RT) \cdot \exp(-\Delta G_D/RT)$$

T = temperature in °K
W^* = the so-called thermodynamic barrier for nucleation, which depends on further parameters.
ΔG_D = the activation energy for transport across the melt-nucleus interface.
The rate of crystal growth, u, is given by:

$$u = a\nu.\exp(-\Delta G'/RT) \cdot (1-\exp(\Delta G/RT))$$

a = interatomic distance
ν = frequency of atomic thermal vibrations
ΔG = the macroscopic decrease in free energy per mole on crystallization
$\Delta G'$ = activation energy for the transfer of material across the liquid-crystal interface.

Several comprehensive reviews by Uhlmann (1977, 1985) and Uhlmann and Yinnon (1983b) cover the theory of nucleation and growth and its applications in considerable detail.

When the parameters needed to apply the above equations are known (or can be estimated with reasonable confidence), satisfactory agreement is usually obtained between calculated and measured values of u and I. The necessary parameters are rarely available for complex commercial compositions. Nevertheless the fundamental work which has been done in this field is of considerable general value, not only in discussing glass formation and devitrification but also in interpreting the behaviour of a wide range of materials in glass technology, which depend on controlled nucleation and growth , for example the glass-ceramics.

One simple observation is worth mentioning, which is qualitatively related to the above theory and which can be of value when searching for regions of glass formation and when developing glass compositions with higher resistance to devitrification. The present author pointed out (Rawson 1956, 1967) that, in many systems, the region

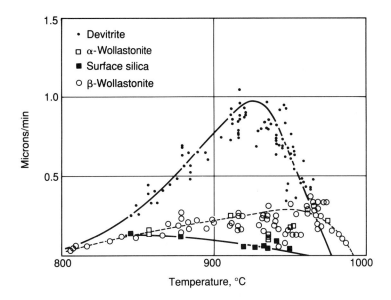

5. Interlaboratory comparison of crystal growth rate measurements on a sheet glass composition *(Physical property committee 1964)*

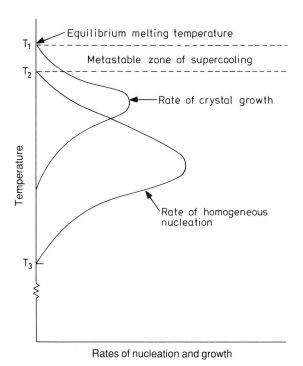

6. Effect of temperature on the rates of nucleation and crystal growth (schematic)

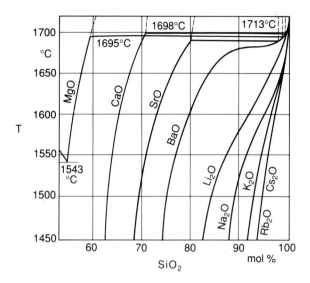

7. The silica-rich end of the phase diagram of alkali oxide-silica and alkaline earth oxide–silica systems *(Levin and Block 1957)*

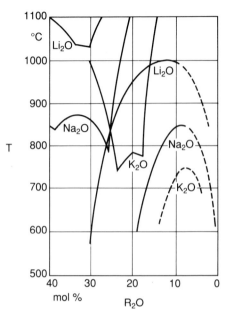

8. The metastable immiscibility regions and the corresponding liquidus curves in the alkali–silicate systems *(immiscibility regions as determined by Moriya et al. 1967)*

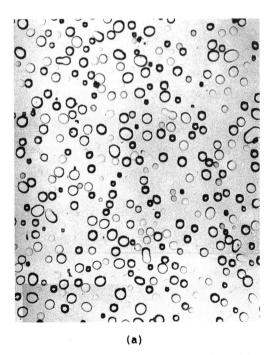

(a)

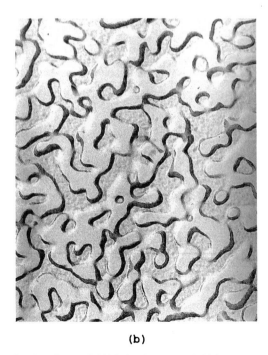

(b)

9. Separation textures: (a) soda-lime-silica glass after heating at 740°C for 7 hr ($\times$ 14 000). (b) sodium borosilicate glass after heating at 700°C for 5 hr ($\times$ 24 000)

of glass formation coincides with a region of the phase diagram where the liquidus temperature is low. Cohen and Turnbull (1961) and others have discussed this liquidus temperature effect in more detail. It has been found to provide a useful basis for the discussion of glass formation in many systems, ranging from commercial silicate glasses to metallic glasses.

1.7. Immiscibility in glasses

Within certain regions of composition and temperature, many two- and multi-component oxide systems are not able to form a single homogeneous liquid phase. Instead they separate into two liquid phases of differing compositions. Amongst oxide melts, immiscibility is known to occur in the alkaline earth borate and silicate systems.

Fig. 7 shows the silica-rich end of the phase diagram of several binary silicate systems. Above a temperature of about 1690°C in the MgO, CaO and SrO systems, the melt consists of two liquid phases of different compositions. The composition range of immiscibility is most extensive in the MgO – and least extensive in the SrO-system. This phenomenon has been well known for many years and the thermodynamic factors that give rise to it are well understood. Such phase separated melts are of no value for making glasses. A material so separated on a macroscopic scale could have no practical application. (Fortunately it is often possible to remove the immiscibility by adding a relatively small percentage of a third component to the composition. See Rawson 1967.)

The thermodynamic factors which cause immiscibility are not confined to the region above the liquidus. It is now known that *sub-liquidus* immiscibility occurs in many systems. This was not fully appreciated until relatively recently, probably because at lower temperatures the immiscibility tends to be on such a fine scale that it is not easily detectable except at the high magnifications attainable with the electron microscope. Fig. 8 shows the regions of sub-liquidus immiscibility in three alkali silicate systems. Immiscibility occurs for temperatures and compositions below the arch-shaped line for the respective system. Strnad (1986) has given a very good account of the scientific and technological aspects of this phenomenon.

Information about the existence and extent of regions of sub-liquidus immiscibility is of considerable importance to glass technologists. Separation, if it occurs, will have some effect on glass properties which one is always trying to control. This is an extra complication making that control slightly more difficult. On the other hand, as will be seen later, some interesting and practically important materials are made which depend on the controlled development of sub-liquidus immiscibility.

Fig. 9 shows two kinds of separation texture which are commonly observed. In one, droplets of one phase are dispersed fairly uniformly in a matrix of the other. In the second texture the two phases interpenetrate, like water in a sponge. Controlled heat treatment causes the structures to become coarser and the separation may then be visible as a cloudiness in the glass.

Chapter 2

Physical and Chemical Properties

2.1. Introduction

This chapter reviews information on the properties of some of the more important glass systems, particular attention being given to the effects of composition and temperature. The systems considered are two- and three- component silicates and borates. It is on these that the more complex commercial glasses are based.

Fuller accounts of the same material have been given by the present author (Rawson 1980) and, in more detail, by Scholze (1988). The extensive compilation of property data on two- and three component oxide glasses by Mazurin *et al* (1983, 1985, 1987) is an invaluable source of information, as is the book by Volf (1961) which should be referred to for compositions and properties of commercial technical glasses.

2.2. Viscosity

2.2.1. Effect of temperature

The way in which the viscosity of a melt varies with temperature is of great importance in all the forming processes to which the melt and (at lower temperatures) the glass is subjected, e.g. the fabrication of glassware and the secondary processes such as tempering, bending, and flame sealing.

In manufacturers' data sheets, certain standard viscosity temperatures or 'fixed points' are often quoted. They are :

> The strain point – the temperature at which the viscosity is $10^{13.5}$ Pa s (or $10^{14.5}$ poise). At this temperature any internal stresses are reduced significantly in a matter of hours.
> The annealing point – the temperature at which the viscosity is 10^{12} Pa s. At this temperature, any internal stresses are reduced to a commercially acceptable level in a matter of minutes.
> The softening point – the temperature at which the viscosity is $10^{6.5}$ Pa s. At this temperature the glass will deform fairly rapidly under its own weight – though this depends upon its shape and dimensions.

The maximum temperature for continuous use corresponds to a viscosity in the range $10^{13.5}$ to $10^{14.5}$ Pa s. This temperature must not be exceeded, for example, during the bake out processes used to degas lamps and TV tubes. Any tempering stresses, whether produced thermally or chemically, will decrease significantly if the glass is held in this range.

(A recent British and corresponding International Standard describes how the viscosity fixed points should be determined – BS 7034 or ISO 7884: 1987).

Higher temperatures are involved in some secondary processes, though only for short periods e.g. in enamelling and in firing conducting silver conductors onto automobile rear windows. The glass must then be well supported to avoid deformation.

The viscosity–temperature curves of a number of technical glasses are shown in Fig. 10. Most commercial glasses fall in this range. The viscosity varies rapidly especially at low temperatures. Therefore viscosity control by control of temperature is very important in all forming processes.

It is clear from the figure that, for some glasses, the viscosity changes more rapidly with temperature than it does for others. Some increase in the slope of the curve can be an advantage in that it allows a process to be operated more quickly, with a consequent increase in output. The compositions of some container glasses have been modified over the years with this end in view. However, with a very steep curve, the container manufacturing process would be impossible to control.

Glasses such as 8363 in the figure, many optical glasses and practically all the 'new' oxide glasses (aluminates, tellurites, vanadates etc.) have such steep viscosity–temperature curves that they cannot be formed into glassware by any of the conventional processes used in the industry. However rough shapes can always be made from any glass by casting, from which the exact shapes required can be made by sawing, grinding, drilling, etc. as required. Some glasses, e.g. vitreous enamels and solder glasses, are required only in powder form so the fabrication problem does not arise. Thus the shape of the viscosity–temperature curve rarely prevents the use of any glass which has technologically useful properties.

An instructive example of this point is given by the barium alumino–borate glasses, which have excellent resistance to attack by sodium vapour at temperatures of the order of 300 °C. They are used in making the discharge tubes of low pressure sodium vapour lamps. It would be impossible (and uneconomic) to make the entire tube from such a glass. The tube is in fact made from a conventional silicate glass which has good forming properties (similar to glass 0080). In the process, the glass flows from the furnace downwards through an orifice. Concentric within the orifice and mounted slightly above it is a small hopper containing the sodium resistant glass. This flows out onto the inner surface of the glass tubing as it forms. Thus the formation of the composite tubing is controlled entirely by the silicate glass and the steep viscosity–temperature curve of the alumino–borate glass is of no significance.

For many oxide glasses, the variation of viscosity, η, with temperature, T, is well represented by the Fulcher equation:

$$\eta = A.\exp\,(B/(T\text{-}T_0)$$

where A, B and T_0 are constants for a particular composition. From the analysis of the results of viscosity measurements on 25 selected compositions within the range of commercial soda–lime–silica glasses, Lakatos *et al.* (1972) derived formulae for calculating the Fulcher constants for any glass composition within that range. This was a valuable contribution, since viscosity measurements, particularly at high temperatures, are very time consuming.

2.2.2. Effects of composition

Fig. 11 compares the viscosity–temperature curves of the three glass-forming oxides SiO_2, B_2O_3 and GeO_2. Silica glass has the highest viscosity at a given temperature of any available commercial glass. This is the main reason for its widespread use in lamp envelopes (high power arc and tungsten halogen) and in radiant heaters. The other glass-forming oxides have no commercial applications as pure oxide glasses.

The addition of small percentages of alkali metal oxides to either SiO_2 or GeO_2 greatly reduces the viscosity. Fig. 12 shows viscosity data for lithium–, sodium – and potassium silicate glasses. Note that the composition is expressed in terms of the concentration of alkali per unit volume. When this is done, the curves for the three systems coincide. This can be understood qualitatively in terms of the structural model described in the previous chapter. One might reasonably expect that the viscosity of a silicate glass would be reduced as the continuity of the silica network is broken down by the alkali oxide, i.e. that the viscosity of the melt would depend primarily on the number of non-bridging oxygens per unit volume.

In the high liquidus temperature alkaline earth silicate ($RO\text{-}SiO_2$) melts, the viscosity at high temperatures also decreases monotonically with increasing RO content, though not so steeply as in the alkali–silicate systems. (Bockris and Lowe 1954; Bockris *et al.* 1955).

The variation of viscosity with composition in ternary systems is more complicated as illustrated in Fig. 13 a and b. Note in particular that the composition effects are quite different at high and low temperatures. Structural interpretation is difficult, partly because in this case the results refer to a weight per cent substitution of silica by the third oxide. Scholze (1988) suggests that various structural factors are involved : (a) the increasing number of non-bridging oxygens, as already discussed (b) changes in the co-ordination number of the R cations with temperature (c) differences between the field strengths of the R cations. Relatively little has been done over the years to develop a structural interpretation of the viscosity–temperature relationships in three-component and more complex silicate glasses.

Another effect of small (<5%) additions of alkali to silica is that the glass devitrifies much more rapidly. When working with fused silica, one must avoid handling the material with bare fingers if the glass is subsequently to be heated to a high temperature. Even the trace amounts of salts left by the fingers on the glass surface will cause local devitrification.

The alkali–silicate glasses have a very poor chemical durability, being rapidly attacked by atmospheric moisture. They are of no commercial value in themselves and are used only as aqueous solutions, which have a wide range of applications (Vail 1952). Adding alkaline earth oxides, especially MgO and CaO, or alumina to a sodium–silicate glass greatly increases its chemical durability, a fact one sees reflected in the commercial 'soda–lime–silica' glass compositions (Table 2, p. 6).

In the alkali–borate systems, the effect of increasing the alkali content is more complicated (Fig. 14). With increasing alkali content, the viscosity decreases monotonically at high temperatures but increases at low

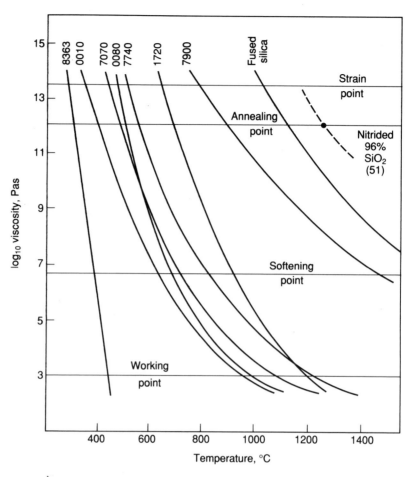

10. Log viscosity vs. temperature curves for some commercial glasses
(Corning Incorporated)

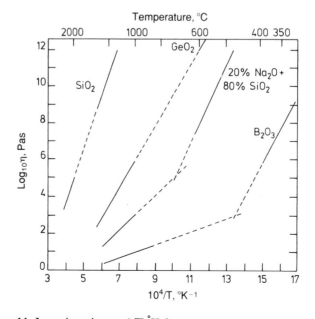

11. Log viscosity vs. 1/T °K for some oxide melts

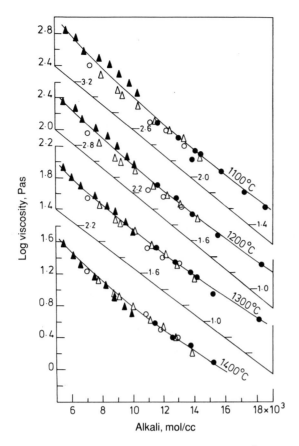

12. Effect of alkali concentration in mol/cm^3 on the viscosity of alkali silicate melts *(Shartsis et al. 1952)*

▲ K$_2$O
○ and △ Na$_2$O
● Li$_2$O

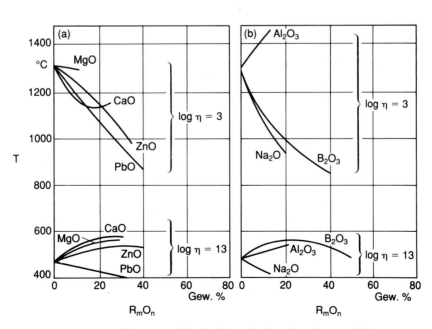

13. Change in temperature corresponding to specified values of viscosity due to weight percent replacement of silica by other oxides
(Gehloff and Thomas 1925, 1926)

temperatures. At intermediate temperatures, there is a viscosity maximum at about 25 mol% alkali. Although it is not possible to explain these trends in detail, the fact that they are very different from those observed in the silicate systems is not surprising in view of the different structural effects caused by the addition of alkali in the two systems (Chapter 1). There is some evidence indicating that for a given composition the proportion of four-coordinated boron is less in the high temperature melt than it is in the glass. Thus the structural change, and the associated property change, depends on temperature as well as on composition.

The alkali–borate glasses are of no practical interest. However the addition of B_2O_3 to alkali–silicate compositions gives the very important group of alkali–borosilicate glasses. The addition makes it possible to produce glasses of low alkali content (<10 %) which (a) are relatively easy to melt (b) have much lower expansion coefficients than the commercial soda–lime–silica glasses and (c) have good chemical durability, especially to attack by acids.

2.3. Thermal expansion

2.3.1. General

The thermal expansion properties of glasses are important in a number of applications and processes:

1. When a glass is fusion sealed to another material. This need arises in the electrical and electronics industries. A range of glasses and related materials, such as glass–ceramics, with carefully controlled expansion properties is required.
2. Stresses produced by temperature gradients are proportional to the expansion coefficient. It determines the effectiveness of a given annealing schedule or thermal tempering schedule. Glasses required to have good thermal shock resistance must have a low expansion coefficient (oven ware and cooker hobs).
3. Some applications require a very high degree of dimensional stability to variations in ambient temperature, e.g. large telescope mirrors and laser gyroscopes.

Fig. 15 shows the expansion curves of a number of commercial glasses and glass–ceramics. Note that the expansion curves for glasses show an increase in slope in the transformation range (see Fig. 1).
Values of expansion coefficient quoted in manufacturers' catalogues are mean values over a specified temperature range. For sealing glasses the range commonly used is 50–350°C. Within the range of glasses which are generally available, vitreous silica has the lowest expansion coefficient ($5 \times 10^{-7}°C^{-1}$). There are however TiO_2–SiO_2 glasses and some glass–ceramics with lower values. Few commercial glasses have expansion coefficients higher than $90 \times 10^{-7}°C^{-1}$, a typical value for container glass and float glass. However there is no problem in obtaining much higher values. Some tellurite glasses for example have expansion coefficients as high as $200 \times 10^{-7}°C^{-1}$.

The familiar borosilicate glass used to make laboratory ware and oven ware has an expansion coefficient of $32 \times 10^{-7}°C^{-1}$ and there are many sealing glasses available with expansion coefficients between that and the value typical of the commercial soda–lime–silicate glasses.

The expansion measurement is usually made on a rod specimen which is under slight axial compression. This explains why published curves often shows a well-defined turnover when the viscosity reaches a value of about $10^{11.5}$ Pa s. The viscosity is then low enough for the specimen to flow under the applied load. The temperature corresponding to this point is referred to as the dilatometric softening point, M_g.

Fig. 16 shows that the expansion behaviour of some glasses – especially borosilicates – is greatly affected by previous heat treatment. It compares curves for the same glass, one a well annealed specimen and the other rapidly cooled. The quenched glass has an initial density which is less than that of the annealed glass. In the transformation range, the structure of the quenched glass begins to change to the more dense configuration characteristic of the annealed material. The densification may be so marked as to outweigh the normal expansion due to the increasing amplitude of thermal vibrations. When the glass has attained the equilibrium configuration, it begins to expand again in the normal way.

2.3.2. Effects of composition

Fig. 17 and 18 show the effects of composition on thermal expansion in the alkali silicate and the alkali–borate systems respectively. As for the viscosity results, the behaviour is quite different in the two systems and for similar reasons.

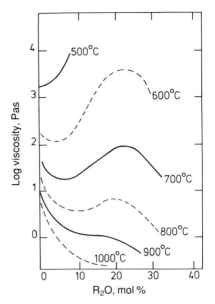

14. Effect of composition and temperature on the viscosity of alkali borate melts
(Shartsis et al. 1953)

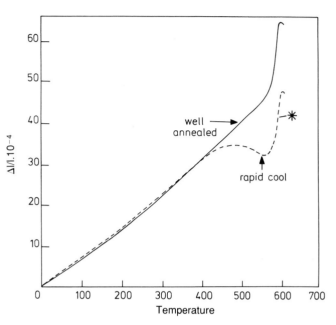

16. Effect of rapid cooling on the thermal expansion of a borosilicate glass
(Turner and Winks 1930)

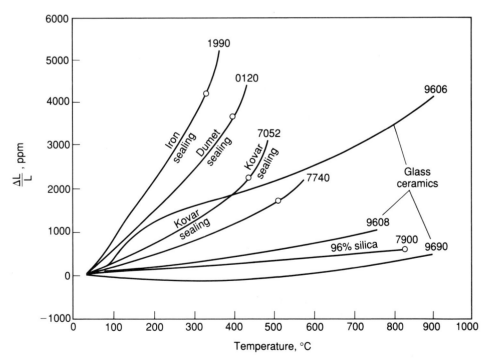

15. Expansion curves of same glasses and glass-ceramics
(Corning Incorporated)

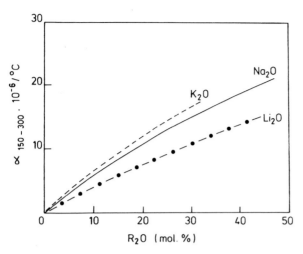

17. Thermal expansion coefficients of glasses in the systems $R_2O–SiO_2$
(Shartsis et al. 1952)

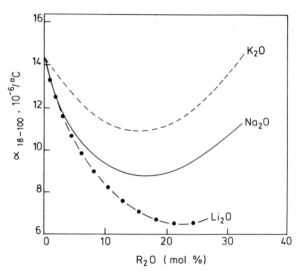

18. Thermal expansion curves of glasses in the systems $R_2O–B_2O_3$
(Various authors)

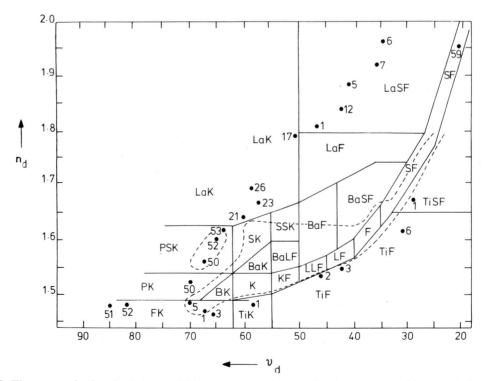

19. The range of refractive index and Abbe number values available in commercial optical glasses
(Schott Glaswerke)

2.4. Optical properties

2.4.1. General

Most optical applications of glasses depend upon two properties:

a. The variation of refractive index with composition and wavelength. The design of high performance lens systems depends on the ready availability of glasses with a wide range of refractive indices and dispersive powers.
b. The variation of transmissivity with wavelength, not only in the visible but also in the infrared and ultraviolet. This is important in signal glasses and in night sights, for which infrared transmitting chalcogenide glasses are used as optical components.

Wide use is made of various surface coating treatments to obtain optical properties not otherwise achievable, such as low emissivity in the far infrared.

More specialised optical applications of glass will be described in later chapters dealing with high transmission glasses for optical communication systems, glass waveguides and photochromic glasses.

2.4.2. Refractive properties (Kreidl and Rood 1965, Parker 1979, Vogel 1985)

Typically, the major manufacturers of optical glasses list about 200 standard glasses in their catalogues. These are high quality materials, of very uniform and reproducible refractive index. The chemical homogeneity is achieved by stirring the melt. The glass must be free from crystalline inclusions and seed and the blocks from which lenses and other components are to be made must be carefully heat treated to ensure that the refractive index is uniform within each block and is reproducible between blocks. For standard glasses, material is supplied having a refractive index which does not differ by more than 0.0002 from the catalogue value. The glass is supplied with the result of an index measurement made to a much higher accuracy than this, since this is necessary for making optical components of the required performance. Within a block, the variation of refractive index can be less than $\pm 1 \times 10^{-6}$ if that degree of homogeneity is necessary.

The general requirement which the optical glass industry has to meet is to supply a range of glasses with which it will be possible to design many types of lens systems and other optical components with the smallest possible aberrations.

Fig. 19 shows a chart of the type commonly used to display the more important properties of commercial optical glasses. The refractive index, n_d, at the wavelength of the yellow helium d line (587.56 nm) is plotted against the Abbe number, ν_d. The Abbe number is a measure of the dispersion and is given by:

$$\nu_d = \frac{n_d - 1}{n_f - n_c}$$

where n_f is the refractive index at 486.13 nm (helium blue), and n_c is the refractive index at 656.27 nm (helium red).

In detailed lens design work, it is often necessary to have information about the partial dispersions in several parts of the spectrum. This is readily available in the manufacturers' catalogues.

Fig. 20 shows, for a number of optical glasses, the variation of refractive index with wavelength. This is mainly due to strong absorption bands which all oxide glasses have in the UV. The dispersion in the visible depends on the strength of these bands and on how close they are to the visible.

Table 3

Compositions of high refractive index glasses for reflective signs

	weight %			
BaO	40.03	51.3	56.0	49.4
TiO$_2$	37.69	34.3	34.0	34.1
Al$_2$O$_3$		3.5	2.0	10.6
B$_2$O$_3$	19.67	10.9	8.0	5.9
ZnO	2.61			
Refractive index	1.915	1.920	1.920	1.900

(See also Tiwari and Das 1972, 1973; Walther and Schäfer 1986)

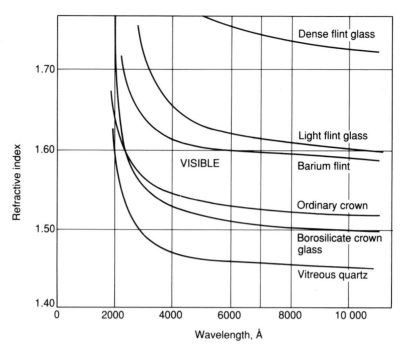

20. Dispersion curves of some optical glasses
(Hagy et al. 1984)

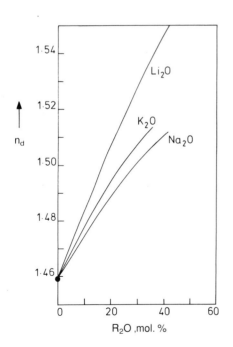

21. Effect of composition on the refractive index of
glasses in the systems $R_2O–SiO_2$
(Vogel and Gerth 1958)

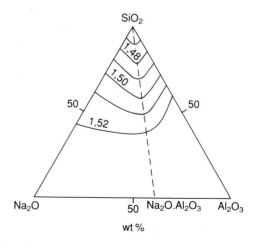

22. Effect of composition on the refractive index of
glasses in the system $Na_2O–Al_2O_3–SiO_2$
(after Schairer and Bowen 1956)

Although oxide glasses are known with optical properties outside the range of the diagram, e.g. with refractive indices greater than 2, only a few have been made commercially as optical glasses. Kreidl (1984) notes that some tellurite optical glasses (which can have refractive indices as high as 2.3) are commercially available.

There is a significant interest in glasses of high refractive index (approaching that of diamond, i.e. 2.42) by manufacturers of reflective screens and road signs. An extensive patent literature exists covering a wide range of compositions. The glass is used in the form of small beads made by spheroidising granules as they fall through a vertical furnace. The rate of cooling of the beads is high, so it is possible to make glasses that would devitrify if melted on a larger scale. Table 3 gives some typical compositions.

Effects of composition
Note that Fig. 20 divides the field of optical glasses into a number of areas according to their properties. Some typical compositions are given in Table 4.

Table 4
Compositions of some optical glasses

	(Hagy *et al.* 1984)									
	SiO_2	B_2O_3	Na_2O	K_2O	CaO	BaO	ZnO	PbO	Al_2O_3	As_2O_3
1	67.1	7.2		16.2	2.0	7.2			0.3	0.1
2	69.6		18.4	11.5					0.3	0.2
3	45.6	4.4		3.9	0.3	32.5	7.9		4.9	0.6
4	36.7	5.9		0.9	0.2	45.1	6.8		3.6	0.8
5	47.5		3.0	9.1	0.3	15.3	8.3	16.3	0.2	0.1
6	52.8			10.1	0.3			36.5	0.2	0.1
7	46.3	5.0		1.1	0.3			47.0	0.2	0.1
8	40.6			7.5	0.2			51.5	0.2	0.1

KEY TO TABLE 4:

		Refractive index			Refractive index
1.	Borosilicate crown	1.5160	5.	Light barium flint	1.5833
2.	Hard crown	1.5175	6.	Light flint	1.5746
3.	Medium barium crown	1.5744	7.	Dense flint	1.6214
4.	Dense barium crown	1.6134	8.	Dense flint	1.6469

(See Vogel 1985 and Deeg 1986 for more extensive information)

They fall into two main classes, the crown glasses with ν_d less than 55 and flint glasses with ν_d greater than that value. Most of the highly refractive glasses contain high percentages of PbO – up to 70 wt-%, whilst low refractive indices are obtained by adding B_2O_3 or a fluoride. BaO is also a common component, increasing the refractive index but to a smaller extent than PbO.

Most of the glasses have properties which lie close to a smooth curve relating refractive index and the Abbe value. However several high refractive index glasses lie well away from this curve. These are members of the group of rare earth borate glasses, discovered by G. W. Morey in 1937. Typical compositions are given in Table 5.

Table 5
Compositions of some rare earth borate optical glasses
(Meinecke, 1959)

Type	ν_d	n_f-n_c	ν_D	B_2O_3	BaO	SrO	La_2O_3	Ta_2O_5	ThO_2
				Weight per cent					
O	1.658	0.011132	58.1	40.0	20.0	20.0	20.0	—	—
N	1.686	0.011828	58.0	40.0	20.0		20.0		20.0
S	1.723	0.01336	54.1	40.0			60.0		
T	1.767	0.01492	51.4	26.0			33.0		41.0
P	1.850	0.02024	42.0	20.0			36.0	28.0	16.0
M	1.898	0.02268	39.6	16.6			37.5	29.2	16.7

The rare earth oxides have the effect of increasing the refractive index. However the fundamental absorption band in the UV is at a significantly shorter wavelength than in the PbO-containing glasses. Consequently the dispersion of these glasses is lower and the Abbe value higher. The use of such 'anomalous' glasses facilitates the design of lenses having high numerical aperture with low spherical aberration.

The refractive index depends partly on the refractivities of the individual ions in the glass and partly on how densely they are packed in the structure. In most two-component oxide glasses the refractive index varies almost linearly with composition as shown by the data for the alkali silicate glasses in Fig. 21.

When a change in the coordination number of an ion occurs, the effect of composition is more complicated. Fig. 22 shows contours of equal refractive index (isofracts) in the Na_2O–Al_2O_3–SiO_2 system.

The isofracts change in direction as they cross the line corresponding to the Na_2O/Al_2O_3 ratio of unity for reasons usually associated with a change in the co-ordination number of the Al^{3+} ions. When the Na_2O/Al_2O_3 ratio is greater than 1 it is possible for all the Al^{3+} ions to be four co-ordinated with oxygen in the glass structure. They substitute for Si^{4+} ions in the network and, to maintain electroneutrality, it is necessary that the glass should also contain a number of Na ions at least equal to the number of network Al^{3+} ions. On the other hand when the Na_2O/Al_2O_3 ratio is less than 1, only a fraction of the Al^{3+} ions can replace the Si^{4+} ions in the network. The rest, it is suggested, are six-coordinated. Because the ionic refractivity of the Al^{3+} ion depends on its co-ordination , the change in coordination number is shown by a change in direction of the isofracts at $Na_2O/Al_2O_3 = 1$. Recent research relating to the structure of aluminosilicate glass suggests that this explanation is not in agreement with all the relevant observations (Smets and Lommen 1981).

2.4.3. Spectral transmission (Weyl 1951; Wong and Angell 1976; Sigel 1977; Bamford 1977)

Many applications of glass depend on the fact that the transmission for electromagnetic radiation varies with wavelength. Oxide glasses have a high transmission in the visible, unless colouring agents are added. Some of the raw materials (sand, limestone and dolomite) are naturally occurring and contain small percentages of colouring impurities, especially iron oxide. This gives a slight tint, blue when the melting conditions are reducing and green when they are oxidising. For a given iron content, the green colour is less noticeable, so it is common practice to add oxidising agents to the glass batch. Another method of achieving a similar effect is to add materials which produce a complementary colour.

Optical communication fibres are made from materials of the highest purity so that absorption losses are reduced to a minimum. Although very low losses have been achieved by crucible melting using highly purified raw materials, most fibres are made by chemical vapour deposition methods using volatile compounds, e.g. $SiCl_4$ previously purified by distillation.

A wide range of coloured glasses is commercially available. Some are made for the domestic market, but others have to satisfy technical requirements for use in optical filters and signal lenses, when it is necessary to be able to produce a particular colour reproducibly.

Coloured glasses are of two types – solution colours and colloidal colours. The colour is not always obtained by making colouring additives to the batch. Colour may also be produced by a secondary, surface treatment process, as in the medieval process for making stained glass or in the modern processes used for making certain types of solar control glass.

Solution colours

The solution colours are produced by adding a small percentage of one or more transition metal or rare earth metal oxides to the batch. A very wide range of colours can be obtained in this way. The colour produced by a particular ion varies considerably, being affected by the glass composition, and (if the ion is multivalent) by the melting atmosphere and by the presence of other multivalent ions in the melt.

The colouring ions are incorporated in the glass structure, each being surrounded by oxygen ions. The colour is due to the excitation of electrons from incomplete shells in the ion to higher energy levels. The energy difference between the levels involved (and hence the colour) is affected by the electric field which the ion experiences from its immediate surroundings. Consequently the colour is sensitive to changes in glass composition. The fundamental aspects of the relation between colour and local structure has been studied in great detail in recent years so that there is now a good understanding of the relationship between colour and structure for many inorganic materials.

The absorption bands associated with transition metal ions are significantly broader than those for the rare earth ions as is shown in Fig. 23. The relevant energy levels in the latter are less perturbed by the influence of the surrounding oxygen ions than in the former and this is one of the reasons why glasses containing various rare earth ions are used in laser glasses.

Fig. 24 shows the absorption spectra of a series of sodium–borate glasses containing Co^{2+}, which gives an intense blue colour in most oxide glasses. However at low alkali contents in this system the colour is pink. As the alkali

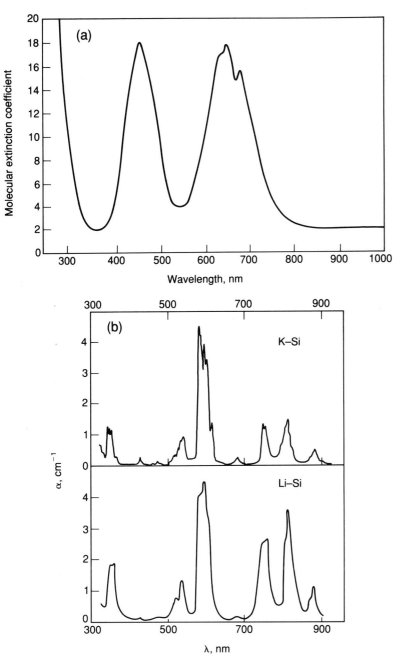

23. Comparison of absorption spectra of (a) a glass containing Cr^{3+} ion *(Bamford 1962)* and (b) glasses containing Nd^{3+} ions *(Patek 1970)*

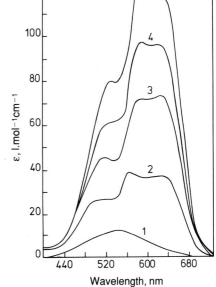

24. The absorption spectra of Na_2O–B_2O_3 glasses containing Co^{2+}. Na_2O contents in mol%: 1, 13%; 2, 22.5%; 3, 26.5%; 4, 30.2%; 5, 33.0% *(Paul and Douglas 1968)*

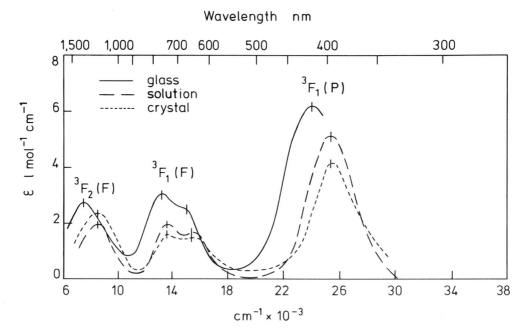

25. Absorption spectra of various media containing Ni^{2+}
(Bates 1962)

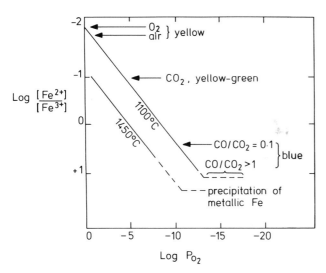

26. The effect of the partial pressure of oxygen on the
$[Fe^{2+}]/[Fe^{3+}]$ ratio in a $Na_2O.2SiO_2$ glass
(Johnston 1964)

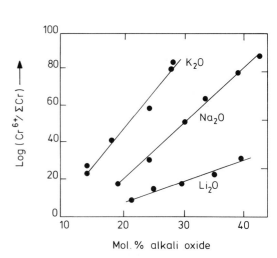

27. The effect of glass composition on the ratio
$(Cr^{6+})/(Cr^{6+}\ Cr^{3+})$ in alkali silicate glasses
(Nath and Douglas 1965)

content is increased, the intensity of absorption in the visible becomes much stronger and the colour changes to the familiar blue. The colour change is due to a change in the co-ordination number of the Co^{2+} ions from 6 to 4.

If the co-ordination number of an ion in quite different solvents is the same, then the absorption spectra are very similar as is shown in Fig. 25 which gives the Ni^{2+} absorption spectra for ions dissolved in an oxide glass, an oxide crystal and an aqueous solution.

Some transition metals are multivalent, the different valence states producing quite different absorption spectra. The redox equilibrium for such an ion dissolved in glass is markedly affected by the glass composition, basic melts being more oxidising. The equilibrium is also affected by the furnace atmosphere.

Fig. 26 shows the effect of temperature and oxygen partial pressure on the Fe^{2+}/Fe^{3+} equilibrium in a sodium silicate melt according to the equation:

$$4Fe^{2+} + O_2 = 4Fe^{2+} + 2O^{2-}$$

Fig. 27 shows the effect of the basicity of the glass on the Cr^{6+}/Cr^{3+} ratio in alkali silicate glasses. As the basicity increases the colour changes from the normal green of the Cr^{3+} ion in silicate glasses to the yellow of hexavalent chromium.

These few examples, selected from a large body of work, are confined to glasses of simple composition. Nevertheless they are of value for indicating those factors which affect the colours due to transition metal ions in more complex compositions. The results are also of value for the scientific information they provide on the chemistry of silicate and borate melts and on the structural effects of changes in glass composition. The latter aspect has been reviewed by Wong and Angell (1976) and Paul (1982).

Colloidal colours

As with the solution colours, the colloidal colours have been known almost since the time when glass was first made, but they have been studied in detail only recently. The major colorants are the elements copper (red), silver (yellow), gold (ruby) and selenium (pink) and the compounds CdS (yellow) and CdSe (red). When used as a body colour, the materials are added to the batch and the colour is produced by controlled heat treatment as the melt cools so producing a colloidal dispersion of particles of controlled size. The chemistry of the melt must also be controlled so that, for example, elemental copper or silver can form during cooling or so that during melting too much Se, CdS or CdSe is not lost by oxidation and volatilisation. Additives such as SnO_2 play an important role, not fully understood, in promoting the formation of the gold ruby.

Copper and silver colloidal colours can also be produced by surface treatment, i.e. by the staining process. The metal compound, usually either the sulphide or chloride, is mixed into a paste with some inert carrier material such as ochre or clay. After applying the wet paste to the glass surface, it is fired for about 30 min at a temperature in the annealing range. The copper or silver ions exchange with sodium ions in the surface layers of the glass. If the glass composition is appropriately chosen, the metal ions in the glass are reduced to the element and the colloidal colour develops during the firing process.

Several workers have studied the kinetics of the base exchange process during staining (Meistring *et al.* 1976) whilst others have shown that the measured absorption spectra of colloidal colours agree very well with those calculated using Mie's theory of scattering and absorption of light by colloidal particles and data for the optical constants of the appropriate elements and compounds (Maurer 1958; Doremus 1964, 1965).

The absorption spectra of these glasses in the visible are of the type shown in Fig. 28 with a high transmission at the red end of the spectrum falling to a low value in the blue.

2.5. Electrical properties (Owen 1963, 1970a, 1970b, 1977, 1985)

2.5.1. General

At room temperature, commercial silicate glasses are very good electrical insulators. One sees glass pressings used for high voltage power-line insulators almost as often as electrical porcelain. Even at 300°C, the glass used in the pinch seal of an incandescent filament lamp provides sufficient insulation between the two lead wires, which are only about 1 cm apart and have 250 volts between them.

The conductivity rises considerably with increasing temperature. In the commercial oxide glasses the conduction is due to the movement of alkali ions under the applied field. As the thermal energy available in the material increases, there is an increasing net drift of the ions in the field direction.

Oxide glasses of high resistivity are normally difficult to melt because of their low alkali content. It is worth noting, however, that alkali free $RO-B_2O_3-Al_2O_3$ glasses can easily be made at relatively low temperatures with

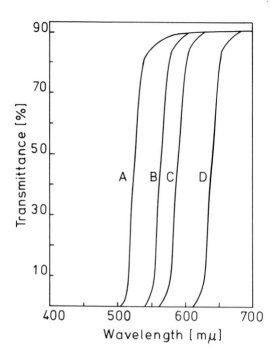

28. Spectral transmission curves of glasses containing:
A - Cds alone; B,C, and D - CdS with increasing amounts of CdSe

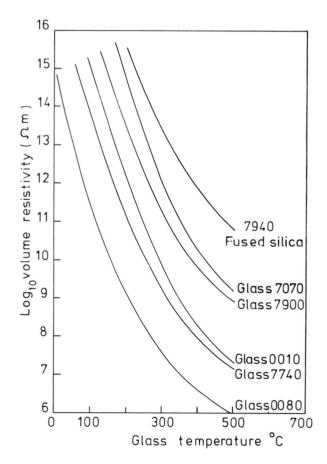

29. Effect of temperature on the dc resistivity of some commercial glasses: 7070 — borosilicate; 7900 — 96% silica glass; 0010 — potash lead silicate; 7740 — borosilicate; 0080 — soda-lime-silicate *(Corning Incorporated)*

room temperature resistivities much higher than that of vitreous silica (Owen 1961).

Fig. 29 shows, for a number of glasses, the variation of resistivity with temperature. At glass-melting temperatures (1500°C), the conductivity is only a few ohm m – low enough to make it possible to melt glass electrically by Joule heating. This is likely to become more common in the future if environmental legislation increases the costs associated with melting using fossil fuels.

Less common but nevertheless interesting examples of the electric heating of glass are electrically-assisted flame sealing and direct high frequency induction melting of glass.

The former technique was widely used to seal the face plate to the cone of black and white TV tubes. The conventional sealing technique is to use flame heating alone, but this has the disadvantage that the glass is heated only from the outside so it takes some time for the inner surface to reach a temperature high enough for the glass to flow sufficiently to make a smooth seal between the two parts. In electrically-assisted sealing, use is made of the fact that gas flames are sufficiently conducting to act as flexible, current-carrying brushes. Each of a pair of opposed burners is connected to a high voltage transformer. The heating of the interior of the glass is now mainly by Joule heating and a much deeper penetration of heat into the glass results.

The second technique was developed during work on melting high purity glass for optical communication fibres. It was found that if the batch is preheated to about 500°C, it is then sufficiently conducting to accept power from a 5 MHz induction coil. Glasses could be melted in an air-cooled high purity silica crucible with no detectable contamination (Scott and Rawson 1973).

Table 6 compares the room temperature resistivities of several different types of glass with typical values for other materials. Most commercial oxide glasses have resistivities greater than 10^{10} ohm m whilst at the other extreme are the metallic glasses with resistivities typical of the metallic alloys. In between are a large number of oxide and chalcogenide semi-conducting glasses.

Table 6

Resistivities of various materials at room temperature

$\log_{10}$ resistivity at room temperature in ohm metre	various materials	glasses	
		probably ionic conductors	probably electronic conductors
25		calcium boro-aluminate glasses	
20		fused silica (synthetic)	
	PTFE		
15		alkali lead silicate glass	arsenic sulphide
	diamond (pure)		
10	nickel oxide (pure) silver bromide	window glass	arsenic telluride
5			CdGeAs$_2$ 90V$_2$O$_5$ • 10P$_2$O$_5$
	silicon (pure)		

Considerable research was carried out on the semiconducting glasses when it seemed likely that they would be used in high speed switching devices. Although the development may not have been as successful as was hoped, the research provided a good understanding of the physics of electronic conduction in amorphous solids, which is proving useful in other contexts, e.g. in xerography (a process which uses semiconducting chalcogenide glasses) and in the development of solar cells based on amorphous silicon.

Currently there is considerable research interest in 'fast ion' conductors – materials with unusually high ionic conductivity, which may be used as electrolyte materials in various kinds of battery. The glasses being studied in this context are of unusual composition containing high percentages of metal halides (Chapter 11).

2.5.2. Alkali silicate glasses

Vitreous silica might be expected to have the highest resistivity of all the oxide glasses in that some grades contain only traces of alkali oxides, present as impurities. However the resistivity is very sensitive to the nature and quantity of impurities. According to Owen and Douglas (1959) the resistivity of the various grades, measured at 300°C, varies over a range of 10^4.

In the binary silicate glasses, the resistivity at a given temperature decreases monotonically with increasing R_2O content (Fig. 30). This is due partly to the increasing concentration of alkali and partly to the decrease in the activation energy (Fig. 31).

2.5.3. The mixed alkali effect

The 'mixed alkali effect' is the most remarkable of all the effects of composition on glass properties and the most difficult to explain. Fig. 31 shows the results of Charles (1965) for three mixed alkali systems. Note that the total alkali content is constant but as one alkali is replaced on an equimolar basis by another the resistivity passes through a pronounced maximum, near the mid-point of the substitution. Charles suggested that the effect may be related to metastable immiscibility. However similar behaviour is encountered in melts at temperatures well above the annealing temperature (Baucke *et al.* 1989). Earlier theories were reviewed by Isard (1969). A more recent theory of Hendrikson and Bray (1972) proposes that it arises from the energy of interaction between dissimilar cations on adjacent sites, which have differing natural vibration frequencies. However in a recent review, Ingram (1987) quotes results which suggest that this theory is not universally applicable.

The mixed alkali effect is exploited in some commercial glasses used in the electrical industry, e.g. in the lead oxide containing glasses used to make pinch seals in tungsten filament lamps. The glass contains both soda and potash in approximately equimolar amounts.

2.5.4. Effect of a third oxide

Below the transformation range, the conductivity of oxide glasses is reduced if relatively immobile divalent ions are introduced into the composition. They occupy a similar position in the glass structure to the alkali ions but are held in place by much stronger ionic forces. They tend to block the movement of the alkali ions through the network. The larger the divalent cation, the larger is the blocking effect (Fig. 33).

Fig. 34 shows the variation of the activation energy for alkali ion conduction in a series of glasses represented by the formula $Na_2O \cdot xAl_2O_3 \cdot 2(4-x) SiO_2$ (Isard 1959). When x is <1, the Al^{3+} ions can occupy network-forming positions, substituting for Si^{4+} ions. Thus at constant alkali content, the number of non-bridging oxygens decreases as the alumina content increases. Isard suggests that this has the effect of expanding the co-ordination shell of oxygen ions surrounding the sodium ions causing the observed decrease in activation energy. At $x = 1$ and above, the Al^{3+} ions enter modifier positions, when they are six-co-ordinated with oxygen. (Note the earlier comments about the interpretation of the refractive index results for the same glass system.)

2.6. Chemical durability

Commercial glasses are required to be resistant to the environment in which they are used or (in the case of bioglasses and controlled release glasses) to react to a controlled and limited extent. For most applications, modern flat glasses and container glass compositions have adequate chemical durability. However even these materials are visibly attacked if they are stored in warm humid conditions, especially if an almost permanent water film condenses on the glass surface. This condition arises, for example, inside a double-glazed unit when the seal has failed. The desiccant soon becomes saturated with water and there is then a situation in which water will condense inside the cavity.

The initial reaction between the water film and the glass results in the extraction of alkali and the pH of the water in contact with the glass rises. This alkaline solution breaks down the silica network resulting in the formation of a white deposit on the glass surface. The outer surface of a window is, of course, even more exposed to water but this is continuously washed away so there is no build up of alkalinity.

Flat glass compositions used in mediaeval times in the windows of churches and cathedrals are far less durable than modern flat glass and, because of the poor control of their compositions, some panes are far less durable than others. The corrosion of many of these windows has been accelerated by their storage during World War II in

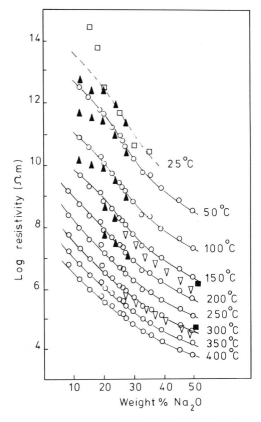

30. Variation of resistivity with composition for glasses in the system Na_2O–SiO_2
(Hughes and Isard 1972)

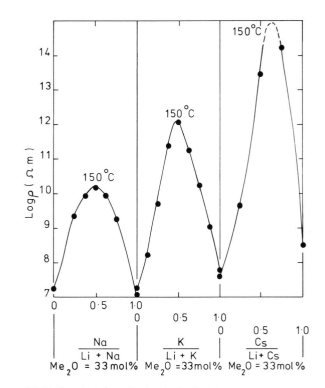

32. DC resistivity of mixed alkali silicate glasses containing 30 mol% R_2O
(Charles 1965)

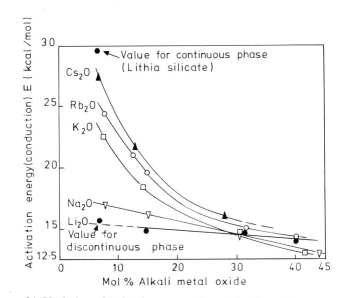

31. Variation of activation energy for conduction with composition in the alkali silicate glass systems
(Charles 1966)

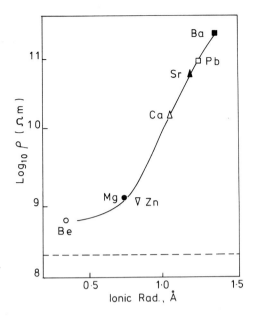

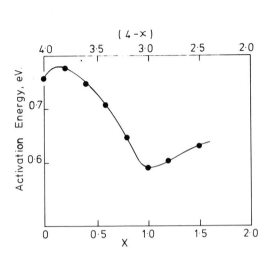

33. The effect of the divalent ion radius on the resistivity of Na_2O–RO–SiO_2 glasses containing 20 mol% R_2O and 20 mol% RO. The dashed line represents the resistivity of a glass of composition 20 mol% R_2O 80 mol% SiO_2
(Owen 1963)

34. Effect of composition on the activation energy for dc conduction in the system $Na_2O.xAl_2O_3.(4-x)SiO_2$
(Isard 1959)

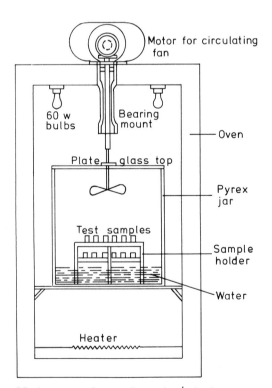

35. Apparatus for accelerated weathering tests
(Simpson 1951)

unsatisfactory environments. A considerable programme of work is in hand in many European countries to develop safe techniques for the restoration and preservation of these windows (Newton 1982, 1985).

Interaction between glass containers and their contents is rarely a problem, but it is essential to ensure that none arises when glass containers are used to hold liquid drugs and blood plasma. A number of durability tests and standards have been published by the ISO and other standards organisations to ensure that the glassware will be satisfactory for that kind of application (Chapter 14). A method sometimes used to increase the durability of the internal surface of pharmaceutical containers is to expose the hot glass to fumes from burning sulphur or some other source of acid gas. This extracts much of the alkali from the glass surface, converting it to sodium sulphate, which is removed by washing.

Resistance of glasses to weathering can be assessed by an accelerated test in which the glass is exposed to warm, humid conditions. One type of apparatus used is shown in Fig. 35 (Simpson 1951). The temperature of the specimen holder is cycled over a 2 hour period at between 50 and 55°C. Water alternately condenses on and evaporates from the glass surface. According to Simpson, this cyclic treatment accelerates the attack. The primary concern is with the visual appearance after the test, but the assessment can be made quantitative, if necessary, by measuring the light scattering produced by the haze on the surface.

2.7. Reactions between glasses and aqueous solutions

Many studies have been made to elucidate the mechanisms of the reactions between silicate glasses and aqueous solutions. Their results have been summarised by Hench (1977); Clark *et al.* (1979); Paul (1982); Newton (1985) and Scholze (1988). The most important factors controlling the rate and the mechanism of the attack are glass composition, the pH of the solution and the temperature.

Fig. 36 shows the effect of pH on the rate of extraction of silica from vitreous silica powder. Above pH 9, the rate greatly increases. In this region the process involves the breakdown of the Si–O network and the formation of soluble silicates in the aqueous phase.

In neutral and acidic solutions, alkali-containing silicate glasses react by a base-exchange process in which Na^+ ions in the glass surface are replaced by H^+, or more probably by H_2O^+ ions. A reaction layer of gradually increasing thickness forms and the amount of alkali extracted increases as $t^{1/2}$. As with vitreous silica, there is a change in the reaction mechanism above a certain value of pH. The rate of extraction of silica increases, due to breakdown of the silicate network but the rate of extraction of alkali decreases. This is probably due to the decrease in the alkali concentration gradient across the reaction layer.

When 'pure' water is in contact with an alkali-containing silicate glass, the water soon becomes alkaline as it accumulates sodium ions extracted from the glass – thus quickly running into the high pH regime of attack. This happens more quickly the higher the ratio SA/V where SA is the surface area of glass exposed to attack and V is the volume of the aqueous phase.

In order to maintain more constant conditions during the extraction experiment, many investigators have used a technique in which the sample of powdered glass is suspended inside a distillation column in which water vapour condenses and flows over the glass powder. Samples of the liquid which has been in contact with a glass are periodically taken for analysis. Experiments of this kind show that initially the alkali extraction increases with $t^{1/2}$ but at longer times the extraction increases linearly with time. It is believed that eventually the surface reaction layer attains a constant thickness determined by a balance between the rate of growth by diffusion of ions through the film and the rate of removal of material by the solution of silica at the interface between the film and the liquid.

Minor constituents in the glass, especially Al_2O_3, ZrO_2 and TiO_2 markedly improve the resistance to attack by aqueous solutions and additives to the solution can also act as corrosion inhibitors.

Hench (1977, 1985a) suggests that the corrosion of glass surfaces can result in the formation of any one of six different situations at the glass–solution interface (Fig. 37). Some are characterised by concentration gradients of glass components in the reaction layer formed on the surface. In addition to the base-exchange and simple solution processes already described, it is possible to have situations in which material extracted is precipitated back onto the glass surface.

Type I is a situation in which the surface has undergone only very slight reaction, with no significant surface layer hydration, e.g. vitreous silica exposed to a neutral solution.

Type II is a surface which has a silica-rich protective film formed by extraction of alkali by base exchange from an alkali-containing glass.

Type III is a more complex structure which sometimes forms on Al_2O_3– or P_2O_5–containing glasses. An

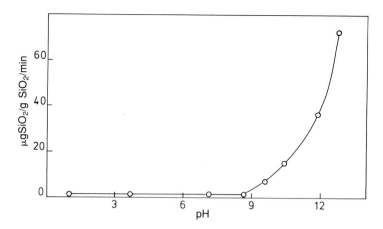

36. Effect of pH on the rate of silica extraction from
vitreous silica powder at 80°C
(Douglas and El-Shamy 1967)

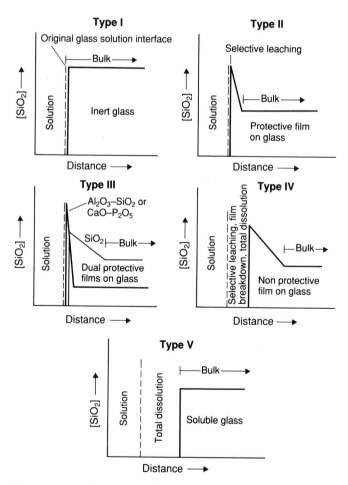

37. Concentration gradients at the glass-solution
interface arising from various reaction mechanisms
(Hench 1985)

aluminosilicate or calcium phosphate layer forms on top of a silica-rich layer. Such layers are very resistant to further attack by both acid and alkaline solutions.

Type IV surfaces have a silica-rich film but the silica concentration is insufficient to protect the glass from rapid attack. Alkali-rich silicate glasses form such surfaces.

Type V is a glass which dissoves congruently, i.e. without significant change in composition. The difference between this and Type I is that a Type V glass dissolves relatively rapidly, e.g. a silicate glass in a strongly alkaline solution.

Clearly the picture is a very complicated one. Which part of the picture is relevant to particular circumstances depends on the glass composition, the composition of the aqueous phase and the conditions of the experiment.

The results of these basic studies on glasses of simple composition are important in many contexts: for the preservation of medieval church window glass, for the use of glasses and glass-ceramics as surgical implants and for the storage of high activity nuclear reactor wastes.

Chapter 3

Strength of Glass and Glassware

3.1. Introduction

The strength of glass is an important factor in many of its applications, as diverse as the glazing of buildings and the manufacture and installation of optical communication fibres. How to attempt to ensure that the strength requirements are met in particular applications will be discussed in later chapters. This one is confined to giving a general account of the more important factors which determine the strength of the material and of the products made from it.

The scientific study of the strength of glass as a material has been and continues to be an important area in glass research. A considerable amount of information is readily available in a number of review articles and symposia proceedings (Zijlstra 1961; Charles 1961; Ernsberger 1973; Wachtman 1974; Freiman 1980; Kurkjian 1985; Rawson 1988).

The theoretical strength of silicate glasses is very high – of the order of 10000 MPa – and strengths approaching this value have been obtained experimentally for material prepared with great care to avoid surface damage and for material from which the surface damaged layer has been removed e.g. by acid etching. It is difficult, if not impossible, to maintain such high strengths because glass surfaces can so easily be damaged in various ways.

All forms of glassware, except glass fibres, contain surface flaws tens of microns deep. They are formed either during the manufacturing process or during subsequent handling of the glass. The relationship between fracture stress, σ_f, and flaw depth, c*, for simple tensile loading is given by the Griffith equation:

$$\sigma_f = (2E\gamma/\pi c^*)^{1/2}$$

where σ_f is the tensile strength
c* is the depth of the deepest flaw in the surface
E is Youngs' modulus of the glass (ca 0.7×10^{11} Pa for commercial silicate glasses)
γ is the fracture surface energy, which is of the order of 4 J m^{-1} for commercial silicate glasses.

Thus the flaw depth corresponding to a fracture stress of 100 MPa (a typical value for several forms of commercial glassware) is 20 μ.

It is of practical importance to understand the mechanisms by which these flaws are produced and the factors that control their growth. A considerable research effort has been devoted to the study of these matters and a corresponding development effort to finding practical ways of minimising damage to glass surfaces and preventing the growth of flaws so that glassware of improved mechanical properties can be made.

Any glass article has a certain distribution of flaws on its surface. This distribution varies considerably from one article to another both with respect to their severity and their position on the surface. Consequently strength measurements made on a batch of apparently identical specimens of glass articles vary considerably, a coefficient of variation of 25% being typical. For practical purposes the minimum fracture stress is particularly important. If the strength values fit a Gaussian distribution, this can be estimated from the mean and standard deviation. However non-Gaussian distributions are not uncommon.

3.2. Revealing the state of damage on a glass surface (The Ernsberger technique)

It is helpful to have a direct visual indication of the intensity and distribution of damage on a glass surface. A simple method of achieving this has been described by Ernsberger (1960, 1962). An ion exchange process is used to develop a controlled tensile stress in the glass surface to a depth of a few microns. After the treatment, flaws which are sufficiently severe to propagate under the stress which has been produced will do so and the cracks can then be seen under a low-power optical microscope. The ion exchange treatment involves immersing the glass in a 40 mol% LiNO$_3$ 60 mol% KNO$_3$ melt for 20 min at 200–250°C. (The optimum time and temperature for a particular glass must be determined by experiment.)

The tensile stress arises because the Li^+ ions are much smaller than the Na^+ ions which they replace in the glass surface. The thin surface layer which has been affected by the base exchange process attempts to shrink but is restrained from doing so by the interior of the glass. Consequently a tensile stress is set up parallel to the surface and equal in all directions. Note that the technique is limited to those glasses containing sodium or larger alkali cations.

In practice one is particularly interested in the more severe flaws. These are detectable using short base exchange times at relatively low temperatures. More extensive base – exchange tends to confuse the picture by revealing the very large number of less severe flaws.

Fig. 38 shows the state of damage on the base of a milk bottle, of the type used in the UK for delivery to the home. The photograph is a plan view of the base. Refractive index matching paint has been used to conceal the flaw pattern on the outer surface of one half of the specimen. The container was made by the press-blow process (Chapter 5). The outer surface is heavily damaged by handling, as one might expect. The inner surface is much less severely damaged except in one area. This heavy damage is caused by metal particles which have been transferred from the plunger to the glass surface.

A detailed study of the strength of float glass surfaces has shown that there is good agreement between the strength results and the area density of flaws revealed by this method (Hamilton and Rawson 1972).

A simpler method of revealing flaws is to etch the glass surface using hydrofluoric acid. More rapid etching of the surface occurs at the flaws. Augustsson *et al.* (1986) have used this technique in a detailed study of the strength of glass containers. They were able to predict the strength properties of a population of containers from measurements of flaw densities and stress distributions calculated by the finite element method.

3.3. Factors affecting the strength of damaged surfaces

3.3.1. Sources of damage (Rawson 1988)

Practically all forms of glassware come into contact with other materials whilst they are being formed, i.e. from the instant at which the glass leaves the furnace. The hot glass may be damaged either by sticking slightly to another surface or by foreign material being transferred to the glass surface. In the glass container forming processes, the likelihood of damage is particularly high since the glass gob, as it is delivered to the machine, strikes a hot mould or plunger at a high speed. In making communication fibres, for which the highest possible strength is required, the fibres may pick up dust from the atmosphere.

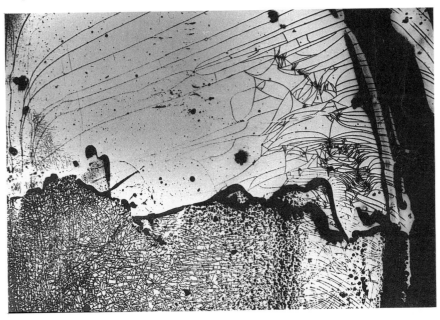

38. Damage on the base of a milk bottle revealed by the Ernsberger technique. The upper half shows the internal damage, the lower half the external damage. Distance across the width of the section is approx -imately 30 mm

The particular examples of damage by contact and by surface contamination which are mentioned below are particularly relevant to glass container manufacturing but similar problems occur in most other glassware manufacturing processes.

Mould materials differ considerably in their tendency to stick to hot glass. Fairbanks (1964) and Fan and Fairbanks (1964) have carried out an interesting laboratory study of the sticking temperatures of many alloys and mould coating materials. They show, using radiotracer methods, that, even with slight sticking, trace quantities of material are transferred from the metal to the glass surface.

It is well known in the container industry that a considerable though localised transfer of metal occurs from the plunger to the glass in the press-blow process with a consequent loss of strength. Transfer can be reduced by the application of semi-permanent coatings to the plunger and mould components, but further development will be required before this problem is completely solved.

In the region between the machine and the lehr, so-called hot end damage is caused by contact between the glassware and the metal components which handle it, e.g. the tongs that transfer the container by its neck from the mould to the dead plate and the guides which direct the container along its path from the machine to the lehr. These sources of damage are gradually being eliminated by replacing uncoated metal with a variety of proprietary composite materials, usually based on graphite or carbon (Vickers 1988). Dust in the air blown onto the base of the container immediately it has been released from the mould may be another source of damage.

Below the annealing temperature the important mechanisms of damage are by friction and by impact. Both on the inspection lines in the glass factory and subsequently on filling lines, there is a considerable amount of rather violent contact between the containers. Frictional damage is considerably reduced by the dual coating process described in Chapter 5.

The most effective way of reducing impact damage is to arrange that the glass surface is under compression. This can be achieved by thermal tempering or by chemical strengthening (Chapters 4 and 5). Fig. 39 shows results of Wiederhorn and Lawn (1979) for the reduction in the strength of flat glass as a function of the energy of impacting particles. Obviously the damage is far less for the tempered glass specimens.

Impact damage by sharp particles may not seem relevant to processes in the glass industry. However there must be many occasions when a glass surface strikes another surface in which a hard particle of dust or other material is embedded. If the glass surface is subsequently stressed fairly uniformly (as in a container under internal pressure) one severe flaw will reduce the strength considerably.

3.3.2. Area under stress

A feature of the type of damage normally encountered on glassware is that the most severe flaws, i.e. those which determine the strength of the weakest articles, are relatively few in number and are some distance apart (several millimetres or even centimetres). Consequently the smaller the area of glass surface which is subjected to stress, the lower is the probability of one of the more severe flaws being located within a highly stressed part of the surface and hence the higher is the fracture stress. The fracture stress also depends on how the method of loading distributes the stress over the glass surface. The following examples illustrate these points.

Table 7 gives modulus of rupture (bending strength) results obtained on 1½ wide strips of plate glass. The first group was broken under three point loading and the other three under four point loading with the inner loading points set at increasing distances, l_2, apart. Note the decrease in fracture stress with increasing area under load.

Table 7
Effect of area under maximum stress on the modulus of rupture
(Kerper and Scuderi 1964)

l_2 mm	Area under maximum stress mm^2	Fracture Stress MPa
Three point		
0	0	124
Four point		
50	1900	121
100	3800	108
200	7600	94

Table 8 gives fracture stresses of float glass surfaces determined by the Hertz fracture method, i.e. by pressing a steel ball against the glass surface until fracture occurs. The decrease in strength with increasing ball size is due to the increasing area of contact between the ball and the glass and hence the increasing area which is subjected to high stress.

Table 8

Effect of ball diameter on the fracture stress measured by the Hertz
method on the two surfaces of float glass
(Hamilton and Rawson 1970)

Ball diameter	Mean fracture stress	
	Tin bath surface	Upper surface
mm	MPa	
6.35	350	600
3.17	480	670
1.59	620	750
0.79	800	900

The marked difference between the strengths of the two surfaces is interesting. It is known that the surfaces differ chemically (Chapter 4). In addition, the tin-bath surface has passed over the lehr rolls which are likely to have produced a considerable amount of damage. It seems likely that the explanation for the difference in strength is mechanical rather than chemical.

3.3.3. Flaw statistics

The area effects and other features of the strength data for brittle solids can be interpreted by the methods of flaw statistics, introduced by Weibull.

In the simple case of a large population of apparently identical glass rods, there will be some stress, σ_u, below which no rod will break if all were subjected to uniaxial tension. Above this stress there will be an increasing number of flaws per unit area, $n(\sigma)$, which are capable of acting as fracture origins at a stress $\leq\sigma$ is the flaw distribution function. It gives a quantitative description of the state of damage of the surface, assuming that this is uniform over the surface.

For uniform tensile loading, a simple analysis leads to the following relationship between the cumulative fracture probability, P, and (through $n(\sigma)$) the applied stress.

$$P = 1-\exp(-n(\sigma) \cdot A)$$

where A is the surface area under stress. It is clear from this equation that the fracture probability increases with increasing area under stress. For this simple example it is easy to calculate from the strength results the variation of $n(\sigma)$ with σ, i.e. to evaluate the flaw distribution function. When the stress is not uniformly distributed over the surface, the analysis of the data is more complicated.

Fig. 40 shows the flaw distribution functions for float glass and plate glass surfaces, derived from Hertz fracture strength measurements (Hamilton and Rawson 1972). The literature now contains many examples of the analysis of glass strength data by Weibull statistics and the use of the parameters so obtained for design purposes (e.g. Gupta 1985, Beason and Morgan 1984).

3.4. Effect on strength of time under load and rate of loading

At room temperature and in atmospheres containing water vapour, the strength of glass decreases the longer the time for which the stress is applied (static fatigue). As Fig. 41 shows, the time effect is temperature dependent and is non-existent at the temperature of liquid nitrogen. For the same basic reason, the strength decreases with decreasing rate of loading – dynamic fatigue (Fig. 42). The static fatigue effect is of considerable technological importance in applications where glass has to withstand relatively high stresses for long periods of time, e.g. in containers for carbonated drinks and in communication fibres. The time effects must also be taken into account in the design of architectural glazing. In some locations, the glass has to withstand severe wind gusts lasting only a few seconds. In others it may have to carry a snow load for several weeks. The values of glass strength used for design purposes in these two situations are very different.

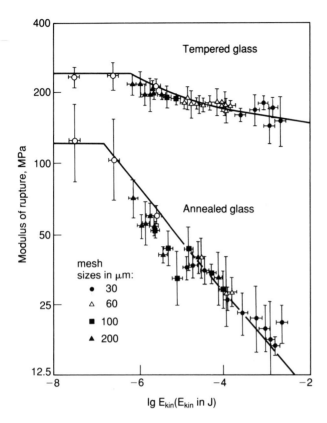

39. Strength degradation of tempered and annealed glass discs after impacting with SiC particles (mesh sizes: 30 to 200 μm). The extent of damage is independent of particle size within the range indicated *(Wiederhorn et al. 1979)*

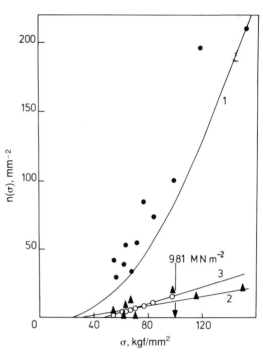

40. Flaw distribution functions for three types of flat glass surface.
1:ground and polished plate. 2: float glass — tin contact
surface. 3: float glass — upper surface.
(Hamilton and Rawson 1972)

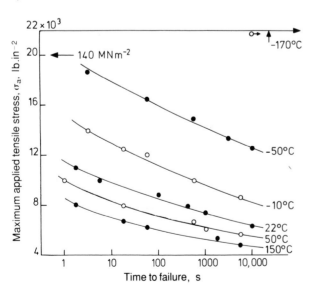

41. Relation between fracture stress and time to failure at various temperatures
(Charles 1958b)

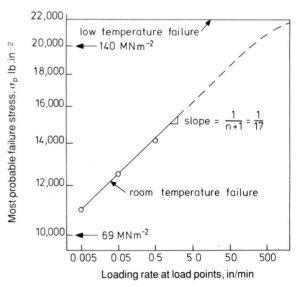

42. Effect of loading rate on fracture stress
(Charles 1958c)

The reason for both fatigue effects is well known. The stress at the tip of a flaw is much higher than the average stress applied to the glass. The high stress accelerates the reaction between water vapour and the glass resulting in growth of the flaw to such an extent that eventually unstable crack growth occurs. The longer the stress is applied or the more slowly it is increased, the greater the reduction in strength caused by water attack.

These effects have been well known since the 1930s and were studied in considerable detail by Charles (1958a, 1958b, 1958c) who provided a theoretical interpretation of his results. In more recent years the emphasis of research has shifted towards detailed studies of crack growth in laboratory specimens, with the use of the methods of fracture mechanics to predict from the crack growth measurements the time-dependent strength properties of glassware.

Fig. 43 shows results of Wiederhorn and Bolz (1970) for the velocity of crack growth in soda–lime–silica glass. The velocity is shown plotted against K_I, the stress intensity factor. At low values of K_I, (region I) the relationship between v and K_I is of the form:

$$v = A \cdot K_I^n$$

n depends on the glass composition and is about 16 for a soda–lime glass. Under conditions of practical interest, region I is of most importance since it covers most of the time which elapses prior to fracture. It is in this region that v is controlled by the reaction between water vapour in the atmosphere and the highly stressed glass at the flaw tip. Note the considerable effect on crack velocity in this region of the partial pressure of water in the surrounding atmosphere.

The curves become almost horizontal at higher values of K_I (region II). It is suggested that in this region, the rate of crack growth is determined not by the reaction rate at the flaw tip but by the rate at which the water vapour can diffuse to the tip along the crack.

In the region of very high crack growth (region III), v is governed by the elastic properties of the glass and no longer by flaw tip reactions.

3.5. Proof testing

Wiederhorn and Evans (1974a, 1974b) have devised a method of proof testing based on the velocity of growth equation valid in regime I. In carrying out a proof test, the component is subjected for a short time to a high stress, which is less, of course, than that which will cause it to break. The time under stress must be so short that no significant crack growth occurs. If the component survives this proof test, in which the applied stress is σ_p, then no flaw in the glass can be larger than a certain size. The minimum time to fracture under the service stress σ_s can then be calculated.

An example of a proof test diagram is shown in Fig. 44. It is for a soda–lime–silica glass immersed in water. If it were required that the minimum service life should be 10^6 seconds under a service stress of 50 MPa, the diagram shows that to guarantee this, the glass must withstand a proof stress of 200 MPa.

The technique has been applied to the testing of space shuttle windows. Its value in the quality assurance of such critical components is obvious. Its application to mass produced glassware may be desirable but would be difficult to engineer.

3.6. The fatigue limit

The existence of the fatigue effects described in the previous two sections raises the possibly worrying question of whether a glass object may fracture eventually even though the stress in it is well below the short time fracture stress. Is there some stress below which no perceptible crack growth can occur?

There appears to be no simple, single answer but a number of comments can be made which may have some relevance :

1. Reducing the stress and hence the stress intensity factor by 10 may be expected to reduce the crack growth velocity by at least 10^{16}. Thus a flaw which takes 1 second to grow to the critical size in a short time strength measurement will take about 10^9 years at the lower stress.
2. A flaw subjected to low stresses will partially heal if it is exposed to a moist atmosphere and so the glass becomes stronger. The extent of this effect depends on the nature of the flaw and on the ambient conditions.
3. Based on a study of crack healing, Stavrinidis and Holloway (1983) conclude that there is in fact a fatigue limit and, for float glass, this is about one seventh of the instantaneous tensile strength.

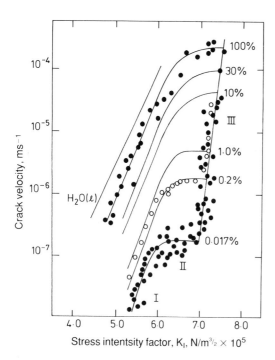

43. Effect of relative humidity on the relationship
between crack velocity and stress intensity factor for
a soda–lime–silica glass
(Wiederhorn 1967)

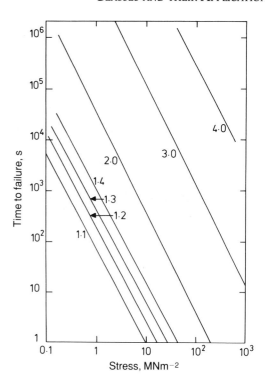

44. Proof test diagram for soda-lime glass
(Wiederhorn and Evans 1974a)

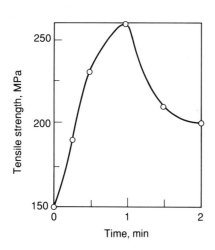

45. Effect on room temperature strength of heating for
various times at 850°C
(Cornellisen and Zijlstra 1961)

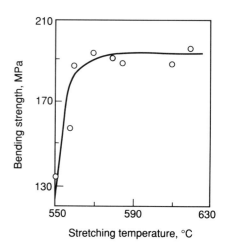

46. Effect on room temperature strength of stretching
the glass at the temperatures shown
(Roeder and Hilpert 1982)

4. There are a number of well known examples which demonstrate the ability of glass to sustain stresses for long periods of time. Thus champagne bottles must survive for several years the tensile stresses caused by the internal gas pressure. The Portland vase, nearly 2000 years old, was made from two different glasses, presumably differing slightly in thermal expansion coefficient. It survived the expansion missmatch stresses, in spite of its mechanically worked surface, until it was broken in the nineteenth century by a deranged visitor to the British Museum.

Thus there are good reasons for believing that, at stresses between one fifth and one tenth of the short time fracture stress, glass will survive for very long times indeed.

3.7. Effect of heat treatment on strength

Surface flaws in glass surfaces can be healed either completely or partially by heating the glass above the transformation range. Thus Hillig (1961) obtained strength values approaching the theoretical value by strongly heating silica glass rods to such an extent that the surface material was removed by volatilisation. Other glasses, e.g. soda–lime glasses cannot be so strongly heated and the increase in strength is less marked. The results in Fig. 45 show an interesting effect for soda–lime glass rods heated for various periods of time at 850°C. After an initial rapid rise, the strength reaches a maximum and then begins to decrease. It is suggested that the decrease is due to slight surface devitrification. It would be interesting to repeat the experiment at temperatures above the liquidus temperature of the glass and preferably in a controlled, dust-free atmosphere.

Fig. 46 shows results of experiments by Roeder and Hilpert (1982) in which rods of soda–lime glass were stretched to twice their initial length at the temperatures specified. This may be related to an effect reported by Budd (1985a) who observed that the internal strength of press-blow containers is greater, the greater the stretch of the parison before blowing.

It is not clear if the strength increase is due to heat treatment alone or whether the stretching of the glass plays a significant role. However it may be relevant in this connection that Berg (1962), by theoretical analysis and by simple model experiments, showed that an elliptical flaw in a sheet of viscous material changes its shape and orientation when it is stretched. If a flaw is initially oriented at right angles to the direction of the uniaxial tensile stress, it first deforms to a circle and then to an elliptical flaw in the stress direction. Such changes in the shape and orientation of flaws in glass would be expected to result in an increase in strength, which may however be non-isotropic.

3.8. Increasing the strength by coating

The point was made earlier that the strength of glass can be significantly increased by prestressing the glass in such a way that its surface is in uniform compression. This can be achieved in several ways – by thermal tempering (Chapter 4), by ion exchange (Chapters 5 and 7), by cladding and by vitreous enamelling. In thermal tempering the compressive surface layer is relatively thick – certainly thicker than the depth of the most severe flaws in the glass surface. The other processes mentioned produce compressive layers with a thickness which may be less than this. Any such process will not then be fully effective. It is pointed out in Chapter 5 that in the base exchange method for strengthening glass containers, there is a small percentage which is not affected by the process.

Before considering the significance of this fact, a few words are needed to explain what is meant by cladding and enamelling. Cladding or flashing of glass is a well-known technique, used since Roman times. By gathering glasses of differing composition from different pots and working the composite gather to shape, an article can be produced in which the main body of glass is coated with a uniformly thick coating of another. The famous Portland Vase in the British Museum was made in this way. The technique can be mechanised and this was first done to manufacture cladded flat glass. If the surface glass has a lower expansion coefficient than the core glass, the surface layer will be in compression when the glass has cooled. Relatively recently, Corning Glass Works introduced a range of domestic glassware, strengthened in this way. A composite sheet is formed continuously by rolling. Shapes are cut from the sheet while it is still hot and these flat shapes are moulded to the forms required – cups, bowls, plates, etc. The surface layer glass in this product is relatively thick so there should be no problem of the kind experienced with ion – exchange strengthening (Dumbaugh et al. 1980).

Vitreous enamelling involves coating the glass with a glass frit of lower expansion than the substrate, followed by heating to fuse the enamel layer. This process also has a long history both in hand working and in machine production. A number of glass compositions suitable for the technique are available. They have a low expansion coefficient and a relatively low fusion temperature (Sozanski and Varshneya 1987). The enamel coatings are rather thin and can be expected similar problems to those encountered with ion-exchange strengthening can be expected.

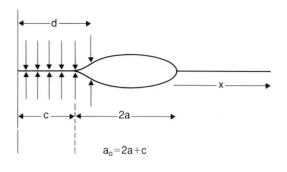

$$a_o = 2a + c$$

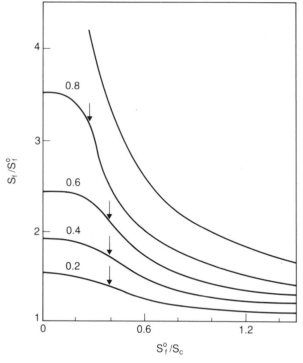

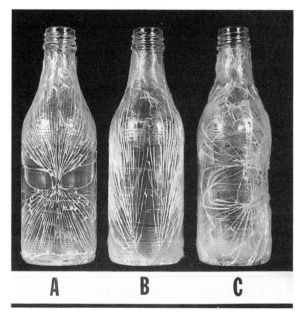

48. Fracture of containers which have been broken by internal presssure (A and B) and by impact, C

47. Strengthening due to surface compression in a semi-infinite plate in terms of the base strength of the material, the magnitude of the compressive stress and the ratio of the compressive zone depth to the surface crack size. The arrows indicate values of S^0_f/S_c below which the crack is partially closed
(Green 1984)

49. Mirror area in the vicinity of a fracture origin
(Orr 1972)

A detailed understanding of the significance of the ratio of coating thickness to flaw depth has been developed only recently. Fig. 47 summarises the results of a fracture mechanics analysis by Green (1984). It shows the factor by which the strength is increased as a function of the surface compressive stress, S_c, and the ratio of the thickness of the coating, d, to the depth of the strength-controlling flaw, $a_0 \cdot S^0_f$ is the strength of the body in the absence of residual stress. S_c is assumed to be uniform through the layer thickness. Note that when d is less than a_0, the increase in strength is limited and an increase by more than a factor of two or three will be difficult to obtain.

3.9. Information from the examination of fracture surfaces (Preston 1931, 1939; Murgatroyd 1942; Orr 1972; Kerkhof 1970; Frechette 1972; Mecholsky *et al.* 1976)

Careful examination of fracture patterns and fracture surfaces can, under appropriate circumstances, be used to determine the cause of fracture, to locate the fracture origin and to estimate the stress at the origin when fracture occurred.

If an abnormally high percentage of glassware breaks on the production line, it is obviously important to find the reason. Also, if during any of the strength tests carried out as part of the quality control procedure, there is a marked change in the pattern of fracture observed, this may indicate that a region of weakness exists in a region of the article where non existed previously. Observation of the change will help to determine the cause.

A crack may propagate in glass over a wide range of velocities up to more than 1000 m/sec for commercial silicate glasses. Very slow cracks, in which the crack tip may take a year to move a few centimetres, are occasionally seen running into the glass from the frame of a window. The crack is driven by the low stresses produced by temperature gradients in the glass and the origin is often a cutting fault. High velocity cracks are more common. These fork at a velocity which depends on the elastic properties of the glass – about 1500 m/sec i.e. the maximum velocity of propagation for soda–lime–silica glasses (Kishimoto *et al.* 1981). The very low speed, low stress fractures are characterised by smooth structureless surfaces; at higher velocities surface markings are seen, which will be described later.

A number of general remarks can be made about crack patterns:

    a. Even though the pattern may be complicated, showing a considerable amount of forking, there is usually only one fracture origin.

    b. The origin is normally at a point on the glass surface, unless there is some gross internal defect, such as a foreign inclusion.

    c. The fracture propagates in a direction perpendicular to that of the maximum tensile stress at the fracture tip.

Fig. 48 shows three glass containers, the first two having been fractured by internal pressure and the third by impact on the side wall. The fracture patterns have been preserved by wrapping each container with transparent adhesive tape before the test. All the fracture patterns show considerable forking. Note that container A shows the classic internal pressure fracture pattern. The initial crack runs, for a very short distance, in an axial direction, i.e. at right angles to the maximum tensile stress which acts in the circumferential direction. In container B, the fracture has started at the heel of the bottle, where damage is particularly likely to occur in the manufacturing process.

The degree of forking in a particular strength test increases with the fracture stress. Thus Frechette and Michalske (1978) found, for each of a number of container designs, a linear relation between the number of branches in the fracture pattern and the hoop stress at failure.

Fig. 49 shows the characteristic appearance of the fracture surface immediately round the origin. There is a smooth, semi-circular or semi- elliptical area which merges into a region with a frosted appearance. The larger the fracture stress the smaller the mirror area, the relationship being:

$$\sigma_f \cdot a^2 = A$$

where σ_f is the fracture stress, a is the dimension of the mirror zone and A is the mirror constant. For a soda–lime–silica glass, A is approximately 2 MPa/m$^{1/2}$ (Freiman 1980).

Examination of the smoother areas of the fracture surfaces remote from the origin may reveal markings similar to those in Fig. 50. These are known as rib marks, each indicating the position of the fracture front at some instant of time. Their curvature indicates the direction of propagation of the crack. This is always away from the centre of curvature i.e. from right to left in the photographs. It is clear from Fig. 50b that the fracture was moving much more rapidly along the bottom surface than along the top surface. This is characteristic of a bending fracture with the lower surface in tension and the upper in compression.

50. Rib marks on fracture surfaces
(Orr 1972)

Chapter 4

Flat Glass and Glazing

4.1. Manufacturing processes

Flat glass has been made through the centuries since Roman times by a number of processes. These include casting, rolling, spinning, blowing and drawing from the surface of the melt in the furnace (Persson 1969, Douglas and Frank 1972; Edge 1984; Pilkington 1976). Now by far the most widely used process is the float process initially brought to success by Pilkington Brothers in 1959. However the rolling process is still used for the manufacture of 'figured plate' glass, the pattern on one of the rolls being moulded into the glass whilst it is still hot. A very similar process is used to make wired glass – one form of safety glass. A small quantity of special glass, e.g. 'antique' glass is made by the old hand-blown cylinder process. The blown cylinder, after being detached from the blowing iron, is cracked along its length and then flattened in a furnace using a rake.

The various applications of flat glass have been reviewed in a number of monographs and articles (Petzold and Marusch 1973; Oliver 1977; Turner 1977; Hammel 1985).

The essential features of the float glass process are shown in Figs. 51a and 51b. A continuous ribbon of glass flows from the furnace onto a shallow bath of molten tin. This is contained in a controlled atmosphere chamber fed with a nitrogen-hydrogen gas mixture to prevent oxidation of the tin. The glass is cooled as it passes along the bath until its viscosity is sufficiently high for it to be lifted from the tin and carried by the transport rolls without deforming under its own weight. A slightly different form of the process was later developed by the Pittsburgh Plate Glass Company. In this variant, the glass is fed from the furnace across the full width of the bath. Also the refractory weir which feeds the glass is in direct contact with the tin.

A simple analysis, originally applied by Langmuir in the context of an oil drop floating on water, has been applied by Charnock (1970) to show that the glass ribbon tends to reach an equilibrium thickness (ca 7mm) as it floats on the tin. This is determined by the densities of the glass and the tin and by the surface energies of the three interfaces: tin–atmosphere, tin–glass and glass–atmosphere.

Thinner glass, down to 2 mm, is made by methods described by Pilkington (1969, 1971) and Edge (1984, 1986). By increasing the speed of the take out rolls (at a given glass flow rate) the glass is stretched longitudinally on the bath. The ribbon is prevented from contracting laterally by a number of water-cooled wheels which bear lightly on the upper surface at the edge of the ribbon. The position along the bath where the stretching occurs is determined by control of the longitudinal temperature distribution in the bath chamber. Thicker glass is made either by reversing the inclination of the wheel axles relative to the centre line of the ribbon so that they exert an inwards force, or by preventing the lateral spread of the glass on the bath by means of water cooled graphite guides. Float glass can be made in thicknesses ranging from 2 to 25 mm.

A limited amount of chemical interaction takes place between the tin and the glass surface in contact with it such that the surface concentration of tin in that surface is of the order of 2%. This falls to a negligible value within a distance of 30 micron from the surface (Sieger 1975, Swift 1984). The presence of tin can be shown, and hence the tin-contact surface identified, by its fluorescence under a UV lamp.

The contamination has no significant effect in the majority of applications. However it is worth noting that, in the manufacture of mirrors by the chemical silvering process, it is much more difficult to silver the tin-contact surface. In any surface treatment of float glass it is wise to check whether or not the process is affected by which side is being treated.

Optical faults in the form of small inclusions, bubbles, lack of flatness and glass inhomogeneity must be kept below specified limits. Bretschneider (1988) has given a brief account of the inspection methods used.

Flat glass is now a very versatile material, partly because of the introduction of several surface treatment processes which have greatly extended its range of application. The following sections will discuss aspects of technology which arise in the more important applications of both uncoated and coated flat glass. They cover mechanical, optical, and thermal aspects.

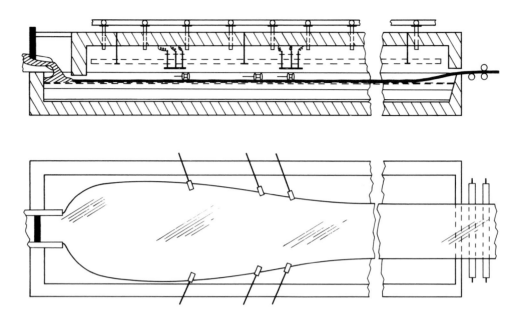

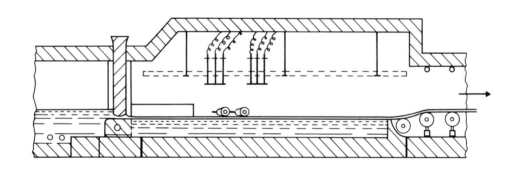

51a. The Pilkington float process
(Edge 1984)

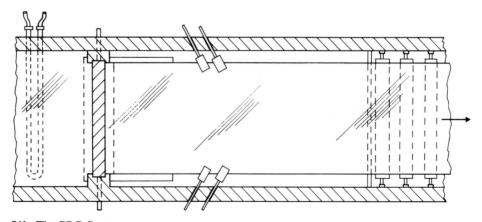

51b. The PPG float process
(Edge 1984)

The main theme of this book is the relationship between the properties of glasses and their applications. The first major application to be considered is that of glazing. In all applications, it must be remembered that success depends on good engineering design and often on the appropriate selection of other materials used in conjunction with the glass. This is certainly true of glazing. The mechanical integrity and weathertightness of any glazing, ranging from a house window to a large section of glass cladding on a modern office building, depend critically on how the glass is fitted into the structure (i.e. on glazing design) and on the correct selection and use of glazing compounds and sealants. Such matters are outside the scope of this book. The reader will find relevant information in a number of publications (e.g. British Standard 6262: 1982; the Glass and Glazing Federation Glazing Manual 1978; Turner 1977). Technical brochures published by the major manufacturers of flat glass and the manufacturers of glazing materials and sealing compounds are also of considerable value.

The following sections will concentrate almost entirely on those aspects of the use of glass in glazing that depend on the properties of the glass itself.

4.2. Strength of glass in glazing

It is an obvious requirement that the glass should not fail under the highest stress which may be encountered. Fracture caused by high wind pressures is not a common occurrence. Nevertheless this possibility has to be considered, especially when dealing with large window areas.

Equations can be found in a number of textbooks for calculating the stresses in uniformly loaded flat plates with specified forms of edge support. Given the strength of the glass, the thickness required to withstand that loading can then be calculated. However the straightforward text book approach does not take into account a number of aspects of the problem: the statistical nature of the strength of the glass and the effect of locality on the probability of encountering strong gusts of wind – to name only two of the important factors. Architects need a scientifically based but uncomplicated way of making decisions on glass thickness which they know has a good foundation in both theory and experience.

The major industrial countries have their own codes of practice covering wind loading design for glazing. Those which apply in the UK are BS 6262: 1982 'British Standard Code of Practice for Glazing in Buildings' and British Standard Code of Practice CP. 3 Chapter V. Part 2: 1972. 'Wind Loads'.

The local wind loading is determined by referring to a map which gives contour lines joining points of equal 'basic wind speed' (BS 6262 and BS 6375. Part 1: 1989). This is defined as the maximum 3 second gust speed likely to be exceeded on average only once in 50 years at 10 m above the ground in open level country. In the UK, the highest basic wind speeds are encountered in north-west Scotland (56 m s^{-1}), whilst in the London area the value is about 38 m s^{-1} (about 85 m hr^{-1}).

In obtaining the design wind pressure from the basic wind speed, a specified allowance is made for the 'ground roughness' around the building under consideration. Four categories of ground roughness are recognised. Very open country is category 1, whilst an area with large and frequent obstructions is category 4. As might be expected, the design wind pressure for a given basic wind speed is considerably higher for a category 1 than for a category 4 environment.

Taking the situation of central London (category 4, basic wind speed 38 m s^{-1}), a table in BS 6262 gives the design wind pressure as 600 N m^{-2}. The final step is to determine the glass thickness required to withstand the design wind pressure. This is done using one of a number of nomograms. Only one is given here, that for single glazing with annealed glass (Fig. 52).

The point of intersection of the horizontal line for the required window area and the vertical line for the determined design wind pressure will lie in or close to one of the diagonal bands corresponding to a particular glass thickness. If the point of intersection lies in an unshaded area, the thickness to use is that for the band immediately above the point. For situations where the intersection point falls within a shaded area, a procedure is defined, based on the aspect ratio of the window (ratio of the greater to the smaller dimension), for opting for either a greater or a lesser thickness.

Table 9 gives recommended maximum areas, obtained in this way, for glass thicknesses of 4 and 6 mm, for two localities and for single and double glazing.

The greater recommended areas for double glazing arise from the fact that the air sealed within the unit transmits a significant fraction of the external load to the inner pane. Thus the stress in each pane is less than for single glazing.

The strength data on which the design charts are based has been summarised by Turner (1977).

Recently a number of papers have been published which examine design for wind loading in more detail. They recognise that the simple engineering formulae used in deriving the BS charts are inaccurate at the large deflections involved. They use the methods of flaw statistics to calculate the fracture probability as a function of the wind load

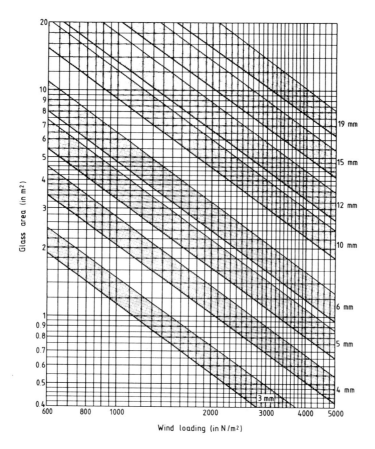

52. Wind loading diagram for annealed glass
(BS 6262: 1982)

Table 9
Maximum recommended areas

Ground roughness: Category 2 – little shelter
Aspect ratio: 1

Location A: Basic wind speed 52 m/sec. Wind pressure 2000 Pa
Location B: Basic wind speed 38 m/sec. Wind pressure 1100 Pa

	A		B	
	Single glazing	Double glazing	Single glazing	Double glazing
4 mm	1.0 m^2	2.0 m^2	1.9 m^2	4.0 m^2
6 mm	2.3 m^2	4.8 m^2	4.8 m^2	9.0 m^2

and they make it possible to calculate the effects of the gust duration and the loss in strength of the glass on long-time exposure.

Minor (1981), Beason and Morgan (1984) and Simiu and Hendrickson (1987) have critically examined the strength data and design methods on which the US design charts are based. Vallabhan (1983) and Tsai and Stewart (1976) have carried out numerical stress analyses to determine the stresses and deflections in glass plates under uniform normal loading. The results have been used by Beason (1986) and Vallabhan and Chou (1986) to calculate the stresses and deflections of the two panes of double glazed units when the outer pane is subjected to wind loading. The problem is to determine the extent to which the air between the panes distributes the stresses between them. Other relevant papers are concerned with the distribution of stresses through the thickness of sheets of laminated glass and how this distribution is affected by the thickness and the elastic properties of the polymer interlayer (Hooper 1973; Behr *et al.* 1986; Vallabhan *et al.* 1987).

4.3. Safety glass

The previous section refers to the higher strength of thermally tempered glass. This is one of several forms of 'safety glass', those most widely used, apart from tempered glass, being laminated and wired glass. The greater safety from the use of these materials arises mainly from the fact that, when the glass breaks, the risk of injury from the fragments is significantly less than when annealed glass is used. With tempered glass the higher fracture stress is an additional safety feature.

The risk of injury depends on the type of accident that may lead to fracture of the glass. Consequently the type of glazing used and the nature of the standard tests used to determine its suitability depend on the nature of the application.

4.3.1. Tempered glass

The tempering process
Flat glass is usually tempered in a horizontal roller hearth furnace. The glass sheets, cut to the required final size and edge-finished, are placed on a horizontal bed of rolls which carry them into the furnace. Within the furnace, the rolls are made from a sintered silica refractory , which reduces the tendency of glass to stick to the rolls and the consequent risk of surface marking. The glass is driven backwards and forwards in the furnace, partly to ensure uniform heating and partly to prevent it sagging under its own weight in the spaces between the rolls. The glass is heated to a temperature of about 650°C. It is then rapidly driven from the furnace into the toughening section, in which two rectangular arrays of blowing heads are mounted, one above and the other below the glass. The jets blow fan air at both glass surfaces until the glass has been cooled to a little above room temperature. Again to ensure uniformity of treatment, the glass is roller-driven backwards and forwards between the blowing heads (McMaster 1984).

Some tempered glass is required in a curved form, e.g. for automobile windows. The equipment used is then rather different. The glass sheets are bent to the curve required at the end of the heating stage after which the glass is immediately toughened. Several different methods have been used to achieve uniform toughening of the curved sheets (Gardon 1980).

Tempered glass is less widely used for automobile windscreens than was the case several years ago. For this application, it is necessary to vary the intensity of tempering over the screen (zone tempering) in such a way that the driver retains some view of the road ahead in the event of a fracture. This is achieved by reducing the intensity of tempering in an area within the driver's line of sight, by placing wire screens between the jets and the glass in this region. The fragments formed in this area are somewhat larger than elsewhere and so a complete 'white-out' of the screen is prevented.

Principles of the tempering process and its control
The basic principles of thermal toughening are easily explained. When the glass reaches its maximum temperature in the furnace, its viscosity is about 10^8 Pa s. At this stage, any stress that may have been previously set up in the glass is very quickly relieved by viscous flow. The intensive cooling to which the glass is subjected when it leaves the furnace produces steep temperature gradients in the direction at right angles to the glass surfaces. Although a parabolic temperature distribution is quickly set up through the glass thickness, this does not produce stresses. The viscosity is still too low. The parabolic profile is maintained throughout most of the cooling but, so long as the shape of this profile does not change, very little stress is produced. It is only as room temperature is approached that the greater part of the tempering stress appears. Removal of the temperature gradients is equivalent to heating the surfaces of the sheet relative to the interior. Consequently the surface layers attempt to expand relative to the interior but are restrained form doing so. The resulting effect is that compressive stresses are set up in the surfaces and balancing tensile stresses in the interior. Thus the stresses are primarily due to the removal of the parabolic temperature distribution. The treatment gives rise to a parabolic stress distribution in the glass as shown in Fig. 53 (p. 50).

Because of the surface compressive stresses, tempered float glass is two to three times stronger than annealed glass. Fracture of glass does not normally occur until a certain surface tensile stress is exceeded. Thus the applied forces must first neutralise the surface compressive stress before any surface *tensile* stress can be set up.

It is not possible to cut or drill glass after tempering. Any operation that penetrates into the central zone of tension will cause fragmentation.

Although the main features of the process can be simply explained in qualitative terms, the complete understanding of the process has not been easily attained. Over the past twenty years a great deal of research, involving measurements and computer calculations, has been carried out to reach the present situation in which there is excellent agreement between measured and computed values of the stresses as they vary during the process and

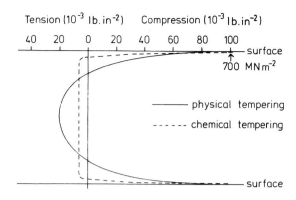

53. Stress distributions in tempered and chemically stengthened glass
(Garfinkel 1969)

also a good quantitative understanding of the factors which determine the final stress distribution. (Excellent summaries of this work have been given by Gardon 1980, 1987.)

The tempering of glass which is more than 3 mm thick is relatively straightforward. Below that thickness, the intensity of cooling required to produce adequate tempering increases significantly and it becomes more difficult to operate the process satisfactorily.

Partly for this reason, development work has been carried out on other methods of tempering (see for example Kieffer and Lindig 1985), some involving the rapid immersion of the hot glass in a liquid bath or in a fluidised powder bed, where high heat exchange rates can be achieved. None of these less common methods appears to have been used in the manufacture of tempered float glass. However they are of interest for more specialised requirements, e.g. for tempering glasses of lower thermal expansion coefficient.

Air jet tempering is also applied to glass components of more complicated shape, e.g. glass pressings used as power line insulators and pressed glass tumblers. For the latter, cooling jets have to be arranged both inside and outside the tumbler. The tempering of glass containers is not considered practicable, partly because the restricted neck opening makes it difficult to insert and remove the toughening jets and to uniformly cool the inner surface. Also modern containers have rather thin walls which are gradually being made thinner (1–2 mm). With such thin glass it is extremely difficult to develop sufficiently high tempering stresses.

As pointed out earlier, the increased strength of thermally tempered glass is not the only reason why it is classified as a safety glass. When properly tempered, most of the glass fractures into a very large number of uniformly sized fragments measuring only a few millimetres across. These are far less dangerous than the large dagger-like fragments which result from the fracture of annealed glass. The formation of the small fragments can be explained either in terms of multiple branching at the tips of rapidly moving cracks (Kishimoto *et al.* 1981) or in terms of the conversion of the high density of strain energy in the glass into surface energy (Barson 1968). The various national quality standards which exist in this field place as much emphasis on the fracture pattern and on the size of the fracture fragments as they do on the strength of the material.

4.3.2. Laminated safety glass

The most common form of laminated safety glass, as used in buildings and in automobiles, consists of two sheets of annealed float glass, bent to shape in automobile windows and windscreens, and bonded together with a tough, tear-resistant polymer – polyvinyl butyral – which adheres well to the glass. Typical thicknesses for a windscreen are 2.3 mm for each of the glass sheets and 0.75 mm for the polyvinyl butyral. The composite sheet is bonded together in an autoclave, typically at 2–6 Kgf/cm^2 and 100°C.

To reduce the risk of laceration in the event of fracture of a laminated windscreen, it is important that the glass fragments are held in place by the polymer and that the screen is not punctured by the impact it receives (Manfre 1985). The most recent step towards reducing laceration injuries from a road accident involving the fracture of a conventional laminated screen is to coat the inner surface with a film of a specially formulated polyurethane (Janderleit 1989).

Considerable efforts have been made to improve the safety qualities of automobile windscreens or to reduce their weight without compromising their performance. Thus Kay (1973) describes a laminated screen in which the outer and inner panes are toughened to different extents. The outer pane is lightly- and the inner pane highly-stressed. If the outer pane is broken by a flying stone, the glass breaks into relatively large pieces so that the driver's vision is not affected. However if the inner sheet is broken by head impact, the highly stressed glass breaks into many small pieces and there is little risk of injury by laceration. Blizard and Howitt (1969/1970) describe a similar development, but using chemically strengthened (Chapter 7) rather than thermally tempered glass.

Glazing having a much higher capacity to absorb impact energy is required in security glazing and in high speed transport, e.g. railway locomotives and aircraft. King and Wright (1987) have given an interesting account of the design and testing of glazing of this kind. The requirements are very demanding and vary according to the application. Thus a common requirement for vehicle security glazing is that it should withstand three rounds of standard NATO 7.62 mm high velocity ammunition. (BS 5051 Part 1: 1973; BS 5051 Part 2: 1979; BS 5375 : 1985.) In addition to stopping the bullets, the glass must not spall on the protected side since the flying fragments would probably injure anyone in that position. In recent designs, a polycarbonate layer is bonded to the inner face of the glazing as an additional safeguard against spalling injuries. Performance requirements for this type of glazing are increasing as more powerful weapons are used in criminal and terrorist attacks. Even now glazing thicknesses of 60 mm are common.

The requirements for aircraft glazing are different. A common hazard is bird impact and a typical requirement is that the glazing should withstand the impact of a 4 lb bird striking the screen at 450 mph. This is tested in a direct way by firing bird carcasses at the screen from a large air gun. In all aircraft windscreen designs, reduction in weight is an important objective. This is being achieved by making use of multilaminates in which the greater part of the structure is a high impact resistant plastic, faced on the outside with a relatively thin layer of glass to provide abrasion resistance. Similarly, in locomotive windscreens, the greater part of the thickness consists of polyvinyl butyral with external and internal facings of glass. It has been shown that the PVB is very effective in absorbing impacts from brake shoes and cinder blocks which may be thrown up from the track or from passing trains.

4.3.3. Standards for safety glass for use in buildings

The standard tests for the quality approval of safety glass differ only slightly from one country to another. They are concerned with ensuring

(1) that the glass is able to withstand a specified level of impact and
(2) that the glass should 'break safely' to minimise the risk of injury if it is broken by human contact.

The tests do differ however according to whether the glass is to be used in buildings or in transport. Also for specialised applications and in research and development, more elaborate tests are used. The testing of aircraft windscreens has just been described. In the automobile industry, tests are carried out, at the development stage, which involve projecting a human dummy against the screen. Such tests assess the risk of injury by measuring the impact forces on the dummy, especially on the head, and by measuring any laceration of the synthetic face of the dummy when the screen breaks.

It is not the intention here to give detailed descriptions of all the safety glass standards, only to give brief summaries which indicate those features of performance which are important.

The British Standard for safety glass for use in buildings (BS 6206: 1981) describes an impact test on a well-supported vertical sheet of glass, the impact being equivalent to that of a small boy running at full speed into a patio door or a large window (impact energy about 600 J). The sheet must either be able to withstand such an impact or it must break safely.

The requirement of breaking safely is determined by setting a limit on the size of the largest fracture fragments. For laminated safety glass a limit is also specified to the size of any hole that may be punched through the sheet.

Safety glass used in the home, e.g. in patio doors, large windows and shower screens is normally tempered. However in high level glazing and roof lights, laminated glass has the advantage that fracture fragments are held in place by the interlayer, so the risk of injury to anyone underneath is less.

4.3.4. Standards for safety glass for use in road transport

The UK standard tests governing the performance of road vehicle safety glass are described in BS AU 178: 1980.

The requirements are more comprehensive than those covering safety glass for use in buildings. As well as specifying strength and fracture behaviour, the standard also defines the required optical properties (especially freedom from optical distortion) stability against discolouration and resistance to moisture attack. Only those aspects which relate to strength and fracture will be summarised here.

Whether the glass is thermally toughened or laminated, the screen or sidelight must fracture safely as a result of either stone or head impact. Impact by a small stone on a laminated screen may cause a fracture, but there will be no danger unless the screen is penetrated. A toughened screen, on the other hand, may shatter and be penetrated . If the screen were uniformly toughened, it would become opaque. To prevent complete loss of vision, zone toughening is used, as described earlier. For a fracture initiated within the visibility zone of a tempered screen the specification requires a maximum fragment size, combined with the retention of adequate visibility. Fragment size limits are also given for side and rear windows.

Resistance to head impact is assessed by dropping a dummy head of specified design onto the glass. The criteria for the performance of laminated screens and windows is that there should be only limited penetration of the glass by the head and that glass fragments should be retained adhering to the interlayer.

It is interesting to note that in recent years, laminated screens have largely replaced tempered screens in British cars whereas they have been standard in American cars for many years. This may reflect the greater difficulty of tempering the more complicated widscreen shapes in modern car designs. Tempered glass is still widely used in side and rear windows.

4.4. Glazing and energy

The basic requirements expected of glazing with regard to light transmission and weathertightness are readily met by taking reasonable care in the design of the glazing and in the selection of materials. However in recent years it has become increasingly possible to meet other requirements which (a) affect the comfort of people living or working in a building and (b) reduce the energy requirements for heating or cooling. A well known example is the development of double glazing to reduce heat losses. Also various types of coloured and surface coated flat glass have become available which reduce glare and control solar gain.

4.4.1. Multiple glazing

Fig. 54a shows a section through the edge seal of a double-glazed unit of conventional construction. The sealing materials are polymers having a low permeability for water vapour and which adhere well to the glass. A variety of sealants is in use. They include polyisobutylene for seal A and polyurethane for seal B. The hollow aluminium spacer bar is partly filled with a molecular sieve desiccant, which has sufficient absorption capacity to take up all the water vapour that is likely to diffuse through the seal during the expected life of the unit (20 years or more).

The diffusion of water is more rapid along the interface between the sealing compounds and the glass. To minimise this, it is necessary to ensure that the glass is clean and dry when the seals are made and that carefully selected sealants are used.

An interesting recent development in edge seal construction involves the use of an extruded polymer strip material which combines the functions of sealant, spacer bar and desiccant (Fig. 54b). The desiccant is dispersed within the polymer and a corrugated aluminium strip, co- extruded with the polymer, serves to maintain the correct spacing. Although service experience with this material is less than with the conventional type of construction, units sealed by this method pass the relevant national standard tests (Hammel 1985).

For satisfactory performance and long service life, it is important that double-glazed units are fitted into frames which have been carefully designed and made from appropriate materials. Failure to take care of installation details may result in premature failure of the edge seals. Adequate information on these matters may be found in BS 6262 and in the Glass and Glazing Federation's Glazing Manual. National standard tests exist to assess the quality of the units and to ensure that they are likely to have an acceptable service life (e.g. BS 5713: 1979).

Thermal conductance. U values

The factors that determine the thermal conductance of a double glazed unit (or any form of glazing) are well understood. The thermal conductance for the central area of the glazing can be calculated by standardised methods, giving results which agree very well with carefully conducted measurements (BS 6993. Part 1: 1989; Guy and Newton 1984).

The factors determining the thermal conductance are shown in Fig. 55. The temperatures T_1 and T_2 are respectively those of the air inside and outside the glazing (*not* the temperatures of the glass surfaces). The thermal conductance, or U value measured in $W\ m^{-2}\ °K^{-1}$, is the heat flow through the glazing per unit area per unit temperature difference $T_1–T_1$.

The figure shows how the temperature falls as one moves outwards through the glazing. The gradients within the glass itself are determined by its thermal conductivity. This has practically the same value for all types of soda –lime flat glass ($1\ W\ m^{-1}\ °K^{-1}$.) The temperature drop across the gas space can be calculated by making use of the results of extensive studies which have been carried out over the years on heat transfer by conduction and convection across gas cavities. It is also necessary to include the contribution of heat transfer by radiation across the cavity, as will be seen later. The temperature gradients in the air close to the inner and outer surfaces of the glazing depend on natural convection heat transfer coefficients, which in building practice are taken to have standard values.

The total thermal conductance of the unit is calculated by treating the problem as one involving a number of well defined individual conductances in series and parallel.

The calculated values are valid only for the central area of the glazing, remote from the frame. Some frame materials, especially aluminium, have a high thermal conductivity and act as a local short circuit for the heat flow.

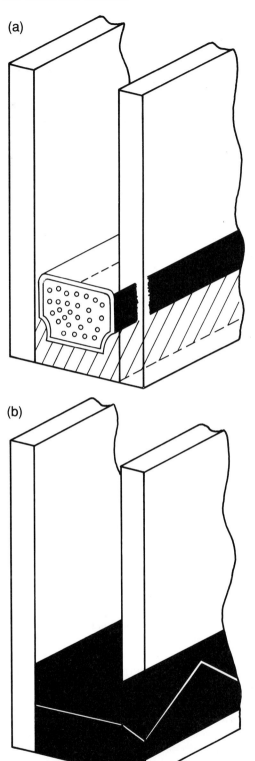

(a)

(b)

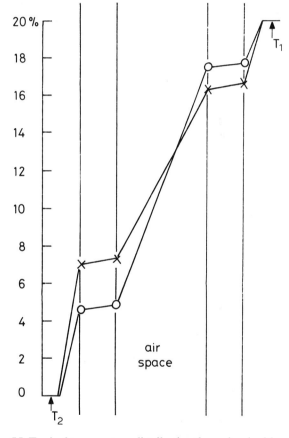

55. Typical temperature distribution through a double-glazing unit, showing thermal resistances

54. Types of edge seal for double-glazing units
(Hammel 1985)
a. Normal seal using two sealing compounds and desiccant contained in spacer bar
b. Sealing strip with co-extruded desiccant and spacer strip

This problem is significantly reduced by breaking the continuity of the aluminium frame internally using polymer strips. Even so, the heat loss through the frame significantly affects the surface temperatures of the adjacent glass (Barnett 1977).

When it is essential to determine the overall conductance of a glazing unit together with its frame, this can be done using standard techniques. The preferred method uses a 'guarded hot box' and is described in BS 874 Part 2: 1987 and BS 6993 Part 2: 1990.

Table 10 gives U values calculated by the method defined in BS 6993. Part 1, for various interpane spacings. The Table also gives surface temperatures of the inner pane, T_4, assuming an inside air temperature of 20°C and an outside air temperature of -10°C. The temperature values clearly show the beneficial effect of double glazing in reducing the tendency for condensation to occur on the inner pane by increasing the inner pane temperature.

Table 10
U values and other thermal properties of double glazing

Glass thickness 6 mm.
The U value for single glazing is 5.35 W $^{-2}$ °K^{-1}
The T_4 temperature using single glazing would be 0.6°C.

Spacing	U value	h_g*	h_r*	T_r*
mm	W m^{-2} °K^{-1}	W m^{-2} °K^{-1}	W m^{-2} °K^{-1}	°C
6	3.1	4.17	3.76	6.8
14	2.7	1.78	3.76	10.3
Using low emissivity coated glass				
6	2.5	4.17	0.57	11.0
14	1.6	1.78	0.57	14.1
Using argon filling. Uncoated glass				
6	2.9	2.8	3.76	9.5
14	2.5	1.2	3.76	10.8
Using argon filling and low emissivity coated glass				
6	2.0	2.8	0.57	12.6
14	1.3	1.2	0.57	15.2

h_g is the heat transfer coefficient across the gas space due to conduction and convection.
h_r is the heat transfer coefficient across the gas space due to radiation.

As is to be expected, the U value decreases with increasing distance between the panes, due to the increasing thermal resistance of the air space. In this range of spacings, the air is stationary so there is no heat transfer by convection. At greater spacings, convection begins and there is little to be gained by increasing the spacing further.

In order to obtain a lower U value at a given spacing, some manufacturers fill their units with gases of low thermal conductivity, the most widely used being argon and sulphur hexafluoride (Hauser 1986). A more commonly used method of reducing the U value is by reducing the radiation heat transfer across the cavity. Glass is commercially available with one surface coated to reduce the room temperature emissivity (low-E glass). Uncoated glass has a room temperature emissivity of 0.83. Coatings reduce this considerably to as little as 0.1 (Hartig *et al.* 1983; Fremaux and Sauvinet 1986). More information on these coatings is given in Chapter 7.

There is no difficulty using current materials in reducing the U value of double glazing to less than one third of the value for single glazing. Even lower values may be obtained using triple glazing, commonly used in countries with particularly cold winters.

Quoting U values for glazing is not particularly helpful unless one can compare them with values for other structural materials. Even the most efficient double glazing does not compare particularly well for U value with that of a brick cavity wall (U<1) and the installation of double glazing in a typical house may not be the most cost-effective way of saving energy. However other factors make double glazing attractive. The table shows that use of double glazing significantly increases the temperature of the inwards facing glass. This makes it much more comfortable to sit near the window on a cold day. Hence a significantly larger part of the room can be effectively used. The comfort benefit can be evaluated quantitatively by a method described by Jennings and Wilberforce (1973) and in CIBSE Guide A1.

The potential contribution of double glazing to energy conservation has been compared with other ways of achieving the same end. There is ample information available for this to be done. As pointed out by Jackson (1987), mandatory standards for home insulation vary considerably from one country to another. He quotes values of a relative energy cost index for houses built to the appropriate building standards in a number of European countries. They vary from 188 for the UK to 100 for Finland. He then shows how the UK figure may be reduced, eventually to 82 by a series of relatively small changes in design and by better use of insulating materials in a fairly conventional house. The most energy effective house was eventually built at Milton Keynes, the only unusual feature of its design being to place 70% of the glazing facing south so as to make best use of the winter solar gain.

The progressive reduction in the energy index is shown below.

Built to current building regulations	188
Arrange 70% of glazing to face south	175
Fit windows with low emissivity coating	159
Double the mandatory roof insulation	154
Double the mandatory wall insulation	133
Introduce ground floor insulation	115

Clearly double glazing is only one feature to be included when designing a new house. It will give worthwhile savings when fitted in an existing house, but other measures may be more effective.

In addition to the advantages associated with reduced heat loss and higher internal glass surface temperatures, there is some benefit from improved sound insulation. However if this is the main purpose of the installation, the design of the glazing unit will be different from that of a unit primarily intended to reduce heat loss (see below).

It has been fairly argued that an energy assessment of glazing based on either calculated or measured U values does not give a proper comparison between glazing and other parts of the building structure (Owens 1982). Glass transmits solar radiation very well in the visible and near infrared. This is absorbed by the contents of the building. The contents radiate heat at much longer wavelengths (ca 15μ), for which the glass is opaque. Thus during daylight hours there can be a considerable energy gain through the glazing and this should be taken into account when assessing the energy saving due to double glazing. A relatively simple way of doing this has been described by Owens (loc cit). Using the appropriate data for solar irradiation (European Solar Atlas 1984) and outside temperatures (Department of Energy. Fuel Efficiency Booklet No. 7), one can readily show that for a south facing window in the south of England, there is a net energy gain if double glazing is used. The effective U value is then negative. This energy capturing possibility has, of course, been exploited in a number of low-energy house designs. However in the average home, the ability to make use of solar gain is not well catered for and it may be wiser to design using the unmodified U values.

4.5. Solar control glass

The increasing use of large areas of glazing in public buildings often leads to problems from glare and excessive heating. With a low sun, glare, in particular, is a common problem. To reduce discomfort from glare and solar gain, the major float glass manufacturers have developed a wide range of solar control glasses of reduced light and solar energy transmission. Two main types are available (a) body tinted glass in which colouring oxides have been incorporated in the glass composition and (b) coated glass which reflects back a significant percentage of the incident radiation.

In buildings with a large area of glazing, the appearance is greatly affected by the type of glass chosen. Thus aesthetic considerations come into account and this is a factor which influences the range of solar glasses which are marketed (Groth and Reichelt 1974; Hussman 1975; Klein 1975; Lampert 1981; Gillery 1982; Geotti-Bianchini and Polato 1983).

Fig. 56 (p. 56) shows how a single sheet of a body-tinted glass redistributes the solar energy which strikes it; 44% is directly transmitted and 5% is reflected. The fraction absorbed heats the glass and consequently the sheet loses heat to the surroundings in both directions. Some of the absorbed heat is re-radiated but the larger part is lost by natural convection to the air in contact with both surfaces. Most is lost to the outer air since the external surface heat transfer coefficient is twice the internal heat transfer coefficient. The net result is that 60% of the incident solar radiation is admitted into the building.

C

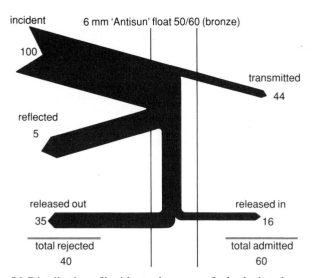

56. Distribution of incident solar energy for body tinted glass
(Pilkington Glass Ltd.)

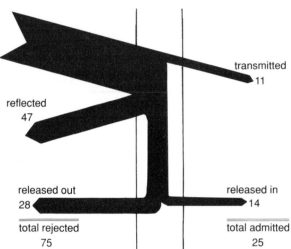

57. Distribution of incident solar energy for coated glass
(Pilkington Glass data)

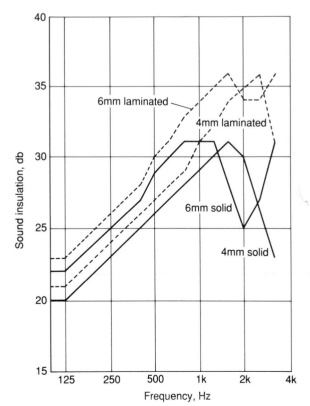

58. Variation of sound insulation with frequency for float glass and laminated glass
(Pilkington Glass Ltd.)

Fig. 57 shows the corresponding values for a glass with a coating of high infrared reflectivity. Such coatings need protection against weathering and abrasion. In single glazing this can be achieved by laminating the coated glass with an uncoated sheet, the coating being sandwiched between the two sheets.

For single sheets, glass is available with a total solar energy transmittance as low as 18% and light transmittance of 14% compared with 83% solar energy transmittance and 87% light transmittance for standard float glass.

Further information about the properties of the commercial products is readily obtainable from the major manufacturers.

In countries such as the UK where air conditioning of cars is uncommon, discomfort due to solar gain is more common when travelling than when at home or at work. A considerable amount of development work is being done to reduce this problem. The requirements to be met are however more complicated and the problems more difficult to solve. A brief account of the various lines of approach has been given by Trier (1988).

The measurements needed for solar gain calculations are relatively straightforward involving determinations of transmittance and reflectance in the relevant wavelength ranges. Complications arise when calculating factors for double and multiple glazing installations since account has to be taken of multiple reflections of radiation between the individual sheets. The basis for such calculations has been described in detail by Rubin (1982a, 1982b) and a proposed standard calculation procedure has been published as ISO/DIS 9050. Again manufacturer's data sheets provide the necessary information for a large number of common designs and glass combinations.

4.6. Thermal stresses

Problems are occasionally encountered with solar control glass installations due to thermal stresses arising from temperature gradients in the glass. The main central area of the glazing is exposed to radiation and heats up. However the glass held in the frame is shielded and so remains cooler. Because of these temperature gradients, tensile stresses are set up at the glass edge in a direction parallel to the edge. These may be so high that the glass fractures. There is a greater risk of this happening with body tinted than with coated glass.

The problem is complicated by the fact that building design features may increase the heating of the exposed glazing. One example of this is when there is a structural member close behind the glazing which reflects a significant amount of the transmitted radiation back towards the glass.

The results of a research study of the problem have been published by Stahn (1977). However the manufacturers publish procedures for estimating the rise of glass temperature, given a particular selection of glass and for particular designs. If calculations using these procedures predict a higher temperature rise than is safe, it is usually possible to avoid the problem by a different selection of glass or a different design (Pilkington Bros. Ltd 'Glass and Thermal Safety').

4.7. Sound insulation properties of windows

In an ordinary home, even the most carefully designed glazing will not solve a serious noise problem. Windows need to be opened from time to time for ventilation. However even in such a building there is some advantage in fitting an appropriate design of double glazing. Near major airports in the UK, grants are available to help residents make such a home improvement. In large buildings which can be air-conditioned, the advantages of fitting windows designed to give good sound insulation are considerable. Valuable information on the principles and practice of sound insulating windows can be found in manufacturer's publications (e.g. 'Glass and Noise Control,' Pilkington Brothers) and in an extensive review of the literature by Marsh (1971a, 1971b, 1971c).

If there is a serious problem, a double glazed unit will be used with an air space of between 150 and 200 mm. To avoid transmission of sound through the frame, each glass pane is mounted in a separate sub-frame, the two being acoustically insulated from one another. A further improvement is obtained by fitting a sound insulating material around the inside surface of the frame (the reveal), but this material must not form a sound transmitting bridge between the two frame halves. The size of the window, and even more the thickness of the glass, have important effects. The sound energy transmitted increases almost linearly with the window size. Generally speaking, increasing the glass thickness improves the sound insulation but the effect is frequency dependent because of resonance effects at certain frequencies which affect the efficiency with which the glass pane vibrations are transmitted between the glass and the enclosed air space. By using panes of differing thicknesses, one avoids a situation in which the optimum coupling frequencies of the two panes are the same.

Table 11 (p. 58) shows the range of sound insulation values obtained with a number of window designs. The values are means over the frequency range 100–3150 Hz.

Fig. 58 gives an indication of the effect of frequency on sound insulation, comparing the spectra of standard float with laminated glass of the same thickness. The use of laminated glass significantly affects the critical frequency of the window and improves the performance at high frequencies.

Table 11
Sound insulation values for various types of window construction

Single glazing	
4 mm glass	21dB
12 mm glass	31dB
Double glazing	
6/12/6	29dB
10/100/6 with internal absorber	42dB
6/200/6 with internal absorber	44dB
6/200/6 no internal absorber	41dB

This chapter has given at least some indication of the large amount of information which is available to ensure the most effective use of flat glass in glazing. This, together with the development of new types of coated glass products has been responsible for a considerable increase in the market for flat glass, especially as a cladding for large corporate buildings, in many cities throughout the world.

Chapter 5

Glass Containers
and other types of Hollow Ware

5.1. Introduction

Glass containers belong to a large and important class of glassware made by pressing and/or blowing hot glass into metal moulds. This also includes articles such as drinking tumblers, beer mugs, lamp lenses and TV tube components. The market for glass containers is, of course, much larger and very different in kind from those for the other types of article which have been listed. The manufacturing processes used have been described in detail in a number of texts (Giegerich and Trier 1969; Doyle 1979; Holscher 1984; Graham and Davey 1984).

In container manufacture cast iron moulds are normally used (Ensor 1978). These may be given a semi-permanent coating to improve the appearance of the glass surface and to prevent excessive damage to it due to contact with the mould (Saeki and Yamada 1982). Other types of hollow ware, of which thin walled drinking glasses and incandescent filament lamp bulbs are examples, are made using quite a different type of mould. This has a porous lining made by baking onto it a paste consisting of a mixture of sawdust or bran and linseed oil. The mould is immersed in water just before the glass is blown into it. The absorbed water is turned to steam when the glass touches the mould and so there is very little contact between the glass and the mould wall. The glass is blown against a cushion of steam and it can be, and normally is, rotated relative to the mould whilst it is blown. As a result, the surface of the article is quite smooth with no evidence of it having been blown in a split mould. Small holes drilled at a number of points through the mould wall allow most of the steam to escape, sufficiently to allow the glass to take up the shape of the mould.

5.2. Glass container fabrication processes

Although the various fabrication processes used in the container industry have been described in considerable detail elsewhere, it may be helpful to give a very brief account here.

In all the hollow-ware processes, with one important exception to be described later, the glass is fed from the furnace to the machine as a sequence of units, called 'gobs'. These are made by shearing a continuous stream of glass flowing through an orifice in the base of the forehearth channel. It is essential that this process is very well controlled. The glass in each gob should be at a uniform temperature throughout and the temperature should not vary from one gob to the next. Temperature variations of a few degrees may cause a significant lack of uniformity, i.e. the ware will vary in weight, and the wall thickness variation will be excessive. It is sometimes possible to see directly that all is not well with the feeding process. The gobs are formed partly by gravity flow and partly by extrusion of the glass through the orifices. They should be perfectly straight and vertical as they form.

If there is a temperature- and hence a viscosity- gradient across the diameter of the gob, it will curve as it emerges from the orifice. This indicates a lack of temperature uniformity of the glass in the forehearth and a need to take corrective action.

Most container machines are of the individual section (IS) type. Several machine sections are mounted on a common frame, side by side, and are controlled from the same electronic timing panel. Each section is fed in turn by the same gob feeder and goes through a similar cycle of operations, but with a phase shift relative to the sections adjacent to it in the feeding cycle. The system is a very flexible one. Thus one unit can be shut down or even removed from the frame, if a fault develops, without interrupting the rest. Machine actions are usually pneumatically driven. Such drives are cheap and robust but at the same time can easily be set to provide some degree of cushioning in the linear or rotary motions which are involved. This feature can be important in reducing forming defects (Hamilton 1977).

All container machines use a two mould cycle (Figs. 59 and 60), the moulds running with surface temperatures of about 400°C (Frank 1961). The gob is first fed to the parison or blank mould which forms the neck region and the finish, including any screw threading. At this stage a hollow shape, called the parison or blank, is formed (either by pressing or blowing), which is then transferred to the second mould for blowing into the final shape.

A considerable amount of heat is extracted from the glass by each mould and this must be removed by the mould cooling – normally by fan air. The distribution of cooling air must be carefully controlled over the surfaces of each mould and must also be appropriately divided between the blank and blow moulds. Poor distribution of cooling results at the best in poor glass distribution and at worst in gross defects on the glass surface, e.g. small surface cracks or crizzles.

An important stage of the process is that when the blank has just been transferred to the blow mould. The glass surface is relatively cold, having just been in contact with the surface of the blank mould. It must be allowed to reheat and the parison allowed to stretch inside the blow mould before the final blow starts. This is critical, especially in making thin walled ware, when it is particularly important to obtain a uniform wall thickness. In recent years there have been significant changes in the techniques for mould cooling and even in machine design, to make it easier to manufacture thin walled ware (Christopher and Murialdo 1977; Heather 1977; Jones and Williams 1983)

There are two types of process, either of which can be operated on the same machine. When making wide necked containers such as jam jars, the parison is made by pressing (Fig. 59). However when making narrow necked ware, both the parison and the final container are often formed by blowing (Fig. 60). Generally speaking, the press-blow process can be operated at a higher speed, mainly because, in the parison mould, heat is extracted from both glass surfaces, i.e. by the mould and by the water-cooled plunger. Also, generally speaking, it is easier to obtain a lighter weight container by press-blow, and one with a more uniform wall thickness.

Not surprisingly, small containers can be made more rapidly than large ones simply because the amount of heat to be extracted is less and the glass must pass through the machine fast enough for it not to set before the shaping process has been completed. Over the years, production speeds per machine have steadily increased, mainly due to the development of techniques to feed simultaneously gobs in groups of two, three and even four to machines which carry the same number of moulds on each mould arm.

It will be appreciated that although sufficiently close dimensional control can be obtained to meet the requirements set by the filling lines in the food and drinks industries, this is not a precision engineering industry. Mould parts wear and this leads to changes in container dimensions and to glass faults, especially in the vicinity of mould joins. Also graphite-containing lubricants, which are from to time applied to the mould parts to alleviate forming problems, tend to build up on the mould surfaces. However, further periods of mould service can be obtained by thorough cleaning of the moulds and replacing lost metal by flame spraying followed by re-machining.

It is likely that in the future the industry will need to invest even more effort into mould coatings than it has so far. This will be necessary to help solve the problem of making containers of increased strength. Especially in the press-blow process, some plunger material is transferred to the inner surface of the container in the form of small particles (Chapter 3). Although the undamaged surface has a strength of about 1000 MPa, the particles reduce this to about 200 MPa (Budd and Cornelius 1976).

Probably the most promising type of coating tried so far in reducing sticking of the glass to the mould during forming (hence minimising damage) is electroless nickel, essentially a nickel–boron alloy (Coney 1978; Frick 1982). However the process has not been widely applied, apparently due to variable adhesion between the coating and the metal. What appears to be needed is a durable coating with little adhesion to hot glass, which will bond firmly to cast iron and will withstand the considerable temperature cycling experienced by the mould surface.

With a total cycle time of the order of 10 seconds and with individual parts of the cycle lasting only two or three seconds, precise control of the machine timing is essential. This is easier to achieve with modern electronic timing equipment than was the case when the pneumatic valves were mechanically operated. The success of the process depends on maintaining a delicate balance and adjustment of heat transfer and mass flow in the glass, phenomena which are linked together through the viscosity–temperature curve.

The basic physics of the process is well understood and a certain amount of data is available on the effects of process variables on glass–to–mould heat fluxes and on glass and mould surface temperatures at various stages in the process (Fellows and Shaw 1978). With that understanding of the principles and with this data, limited though it is, it has been possible to develop computer models which are of considerable help in predicting what will happen if any process parameter is changed (Owens, Williams and Sa 1986).

5.3. Pressing to close dimensional tolerances

Other moulded ware is made to closer dimensional tolerances than those required in the container industry. Valstar (1979) has given an interesting account of the pressing of TV screens. During process development, considerable

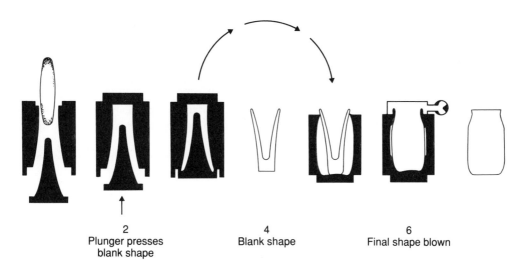

2
Plunger presses
blank shape

4
Blank shape

6
Final shape blown

1
Gob dropped
into blank mould

3
Blank pressed

5
Blank transferred
to blow mould

7
Finished jar

59. The press-blow process
(Moody 1977)

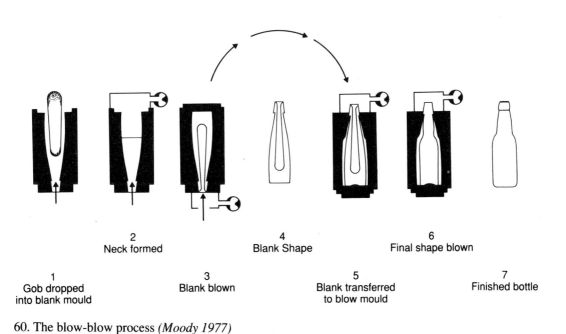

2
Neck formed

4
Blank Shape

6
Final shape blown

1
Gob dropped
into blank mould

3
Blank blown

5
Blank transferred
to blow mould

7
Finished bottle

60. The blow-blow process *(Moody 1977)*

use was made of measurements of glass and mould temperatures to obtain the required screen quality. One measure of the dimensional control is the saggital height, which must not deviate by more than 0.3 mm from a specified value. The saggital height is the height of the centre of the screen above a reference plane when the screen is supported horizontally. Bearing in mind the large weight of the pressing (ca 12 Kg), this is a considerable achievement. Valstar states that the gob weight is controlled to ± 50 g, the gob temperature to $\pm 0.25°C$ and the mid-mould temperature to $\pm 10°C$.

When making very small components, much closer dimensional tolerances can be met. Thus Garfinkel (1986) describes the manufacture of moulded aspheric lenses in which the lens replicates the mould to within $\lambda/10$ or 0.06 μm peak-to-peak.

Graham and Davey (1984) quote dimensional tolerances for glass containers. These depend on the size, but a typical diameter tolerance for a medium size container is ± 1.5 mm.

5.4. Paste – mould machines

The most important and most innovative of the paste-mould machines is the Corning ribbon machine introduced by Corning Glass Works in the mid-1930s for the manufacture of bulbs for tungsten filament lamps. The glass is not fed to the moulds as a sequence of gobs (Fig. 61). A continuous ribbon of hot glass is formed by passing it between two rolls, one, the pocket roll, having recesses machined in one surface. Thus the ribbon has circular, relatively thick sections formed along its length. The machine forms each of these into a bulb. At first the thick sections register over holes in a segmented metal conveyor and the glass begins to sag through the holes under its own weight. The figure shows how blowing heads on another endless belt are brought down to register over the holes in the support belt and how the paste moulds are brought up to close around the glass at the appropriate point. Here they have just emerged from a trough where they have been saturated with water. After the moulds have closed they rotate about the glass. Finally they open and the bulbs are mechanically stripped from the ribbon.

Very high production rates are possible – about 1 000 000 per day for the normal domestic size of lamp bulb. The same machine can make other types of hollowware, for example the inner and outer sections of vacuum flasks.

For a number of years the Owens Illinois Company put a considerable amount of effort into an attempt to make glass containers by this process (Anon 1968). Provided that sufficient strength could be obtained, the idea would still seem to be attractive. One problem could not be solved and that appears to have ended the development. It was not found possible to produce an absolutely flat cut at the neck either mechanically or thermally – an absolutely necessary requirement if a liquid- and pressure- tight cap is to be fitted on the neck.

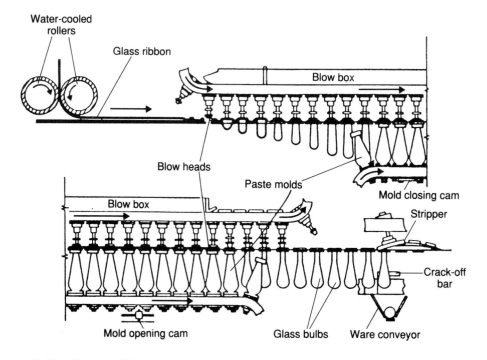

61. The Corning ribbon machine
(Hutchins and Harrington 1966)

5.5. Coatings

If glass is to retain its present market share as a container material, it will be necessary to significantly increase the strength of glass containers so as to make further lightweighting possible (LaCourse, Varshneya and Alderson 1985). It has been suggested that a strength increase by a factor of five is necessary. In recent years a considerable amount of research and development has been carried out on the problem within the industry and in research groups supported by it. It is necessary to prevent damage to the glass surface throughout its useful life and to develop processes that will give a further increase in strength to a relatively undamaged surface.

In the previous section reference was made to the need to reduce damage caused by contact between the hot glass and the mould. Shortly after the container leaves the mould, significant handling damage occurs while the glass surface is still above the annealing temperature, before it enters the lehr. This can be produced by take-out tongs, which remove the container from the blow mould and by various guide bars. A range of temperature resistant materials is now available, used sometimes as facing inserts on metal components, which significantly reduce this hot end damage (Anon 1979; Budd *et al.* 1980; Stewart and Spear 1984 ; Bourne *et al.* 1984; Budd 1985b; Vickers 1988).

Later, both in the glass plant and on filling lines, the containers experience a considerable amount of maltreatment. Impacts and sliding contact between the containers themselves and between the containers and metal parts of the handling machinery produce strength-reducing flaws in the glass surface. Line breakages are costly since they interrupt production. According to Jackson and Ford (1981), the fracture of only 1 container in 10 000 results in a 60% downtime. When a fracture occurs on a filling line, it has to be cleared of broken glass for some distance on each side of the broken container to ensure that none of the nearby unbroken containers carry on through the process containing fragments of glass.

To reduce the damage and loss in strength by impact and sliding contact, nearly all containers are now given a surface treatment which produces a low friction, wear resistant surface (Rawson and Geotti-Bianchini 1985; Budd 1988). Just before the containers pass into the lehr and when the glass-surface temperature is still above 500 °C, they pass through a hood through which a stream of air flows, carrying a controlled concentration of a volatile tin compound – usually $SnCl_4$ (Jackson and Ford 1981). This results in the formation on the glass surface of a layer of tin oxide about 10 nm thick. At the cold end of the lehr, the containers are given a coating of an organic lubricant either by spraying or by vapour treatment. Materials used include polyethylene, polyethylene glycol and oleic acid.

It is well known that many organic compounds, especially long chain aliphatic compounds, greatly reduce the coefficient of friction between glass surfaces, from values approaching 1 for a very clean glass surface to values as low as 0.05. This reduces the contact stresses when one glass surface slides across another. However for a low friction coating to be practically useful, it must also be wear resistant.

The low coefficient of friction must be maintained even when one container is rubbed with considerable force against another. One way of achieving this is to use a lubricant which bonds strongly to the glass surface, such as octadecylamine. This gives an identical coefficient of friction to, but a much greater wear resistance than, stearic acid. The only difference between the two materials is that the former has an amine end group whilst the latter has a carboxyl end group (Jackson, Rand and Rawson 1981). This illustrates the importance of surface bonding of the lubricant in determining wear resistance. In the dual coating technique described, the stronger bonding and increased wear resistance is provided by the tin oxide coating, which is effective when used with a wide range of organic lubricants.

The following results of Southwick *et al.* (1981) for the bursting pressure of containers after abrasion demonstrate the effectiveness of the coatings in reducing damage (Table 12).

Table 12
Strength of coated containers after abrasion

	Relative Bursting Pressure
Uncoated	1.0
Heavy hot end coating only	2.44
Cold end coating only	2.88
Heavy hot end and cold end coating	2.53
Medium hot end and cold end coating	3.00

Smay (1985) has published a valuable review of the literature on the interactions between organic compounds and glass surfaces, with particular reference to their role in damage resistant coatings.

5.6. Polymer coatings

The industry uses polymer coatings for various reasons, e.g. to reduce damage, to retain glass fragments in the event of fracture and to provide decoration and labelling. These coatings are relatively thick (>20 μm) and may reduce damage by cushioning the forces produced when containers come into contact (Brungs 1988). Fragment retention requires the use of tough, tear resistant polymers. Widespread use has also been made of plastic sleeves which, in addition to reducing damage and retaining fragments, carry the decoration and labelling required (Schneider 1977; Kawabata *et al.* 1981; Graham 1986).

It is worth noting that polymer coatings are now widely used to maintain the high strength of silica optical communication fibres. Recent work suggests that the knowledge and experienced gained in this field may result in improved protective coatings for containers (Anon 1988a).

5.7. Glass containers of increased strength

Few attempts to significantly increase the strength of glass containers have reached the production stage. Two are well described in the literature (Poole and Snyder 1975; Watanabe *et al.* 1980; Ono 1981). Both processes involve the development of a compressive layer in the glass surface by an ion-exchange treatment (Chapter 7). A solution containing KNO_3 or K_3PO_4 is applied to the inside and outside glass surfaces after which the container is heated for about 20min. at a temperature just below the strain point. Replacement of sodium ions in the glass near the surface by the larger potassium ions from the salt results in a compressive stress being produced of about 300 MPa. Practically worthwhile increases in strength were obtained – for example a 50–70 % increase in bursting strength.

The improvement that is achievable is determined by the thickness of the compression layer, which is about 20 μm. The most severe flaws which exist are considerably deeper than this – 30 to 40 μm. The coating has no beneficial effect on these and may indeed cause them to propagate at a lower applied stress than in an untreated container (see Chapter 3). Thus although the process succeeded in its main aim, it was found necessary to apply 100 % proof testing to the containers to ensure that the few weak individuals were eliminated.

Bursting pressure results reported by Guillemet (1985), but without detailed discussion, are shown in Fig. 62. They show that a considerable increase in strength can be achieved by base-exchange strengthening combined with the dual coating process described in the previous section. However, Guillemet comments that it is still necessary to screen out those containers which are unaffected by the base exchange process.

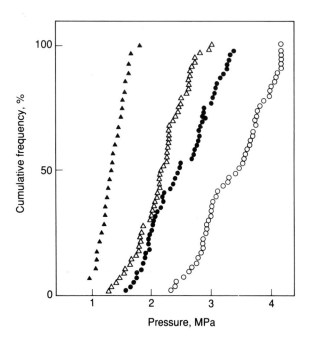

62. Bursting pressures of containers strengthened in various ways

▲ standard bottles

△ with SnO_2 treatment

● chemically strengthened

○ SnO_2 + chemically strengthened

(Guillemet 1985)

Better results might be achieved by using a different glass composition giving a greater depth of base exchange but the scope for improvement in that direction is limited.

5.8. Quality control (Lomax 1983; Wasylyk 1986)

Containers are subjected by the manufacturer to an extensive quality control procedure, involving many measurements and observations.

Typically they include:

1. Container weight
2. Container capacity
3. Dimensions – especially height and neck dimensions
4. Glass quality – absence of foreign inclusions and seed
5. Glass surface quality
6. State of annealing
7. Consistency of colour
8. Surface cracks – checks – especially in the neck
9. Strength – bursting pressure, thermal shock resistance, impact strength, vertical load strength – as appropriate.

Some years ago, practically all these tests were carried out off-line on samples taken from production. Now many are made automatically on-line with some degree of automatic feed-back control. Thus a signal from a weight sensor can adjust the feeder mechanism to maintain constant gob weight. The use of on-line optical sensors to detect checks (small cracks) in the finish is commonplace. Each container is brought to a station where it is rotated. Narrow beams of light are directed at the finish and nearby photocells detect light reflected from any checks in this region. Any defective container is automatically rejected. Some dimensional checking is also done automatically and a reciprocating plunger moving down inside the neck ensures that this is clear. In some plants, all containers are subjected to a squeezing force by passing them between rotating, rubber covered rollers. This eliminates low strength containers by breaking them.

A fairly recent development is the provision of means for optically reading a code moulded onto the container base, which identifies a container as coming from one particular section of the machine (Wasylyk and Southwick 1985). This greatly facilitates the analysis of data coming from the quality measuring sensors on the machine so that both management and operators can easily see how the quality from each of the sections compares.

The strength measurements, which form part of the quality control procedure, are especially important to the bottler and to the final customer. At present they are normally carried out on samples off-line. A good review of the measurement methods and of the standards in use has been given by Lomax (1983). The various tests simulate the loading conditions which the containers experience at some stage in service. The actual value of the fracture stress of the glass itself is not determined.

The internal bursting pressure is carried out by the method specified in ASTM Standard C147–76. The container is filled with water, held suspended by the neck in a water-tight collar and the water pressure is increased at a controlled rate. The test finishes when the container breaks or when a preset upper pressure is reached. The bursting pressure or a failure to break is recorded.

The ASTM standard specifies the method of test but not the strength requirements. The latter are discussed briefly both by Lomax (1983) and by Southwick and Wasylyk (1986). For some time the container industry has worked to voluntary standards agreed between the container manufacturing and bottling industries, but often with an element of independent endorsement. For example, in the USA the Voluntary Products Standard for soft drink bottles is administered by the National Bureau of Standards whilst the European Standard is covered by the data sheet DT5 of CE.TIE. The minimum bursting pressures required are in the range 10 to 15 bars, according to container type and method of filling. Some apparent differences between the specified minimum bursting pressures of the various standards are probably due to differences in values of the acceptable quality levels rather than to any real differences between bursting strengths required in the various countries.

Recent British Standards (BS 6118: 1981 and 6119: 1981) specify minimum pressure strengths for, respectively, multitrip beer or cider bottles and bottles for carbonated soft drinks. The values are similar to those in the voluntary standards.

Wasylyk and Southwick (1985) have recently described commercially available equipment which automatically samples the line production and then subjects each container in the sample to the standard pressure test.

A thermal shock test must be applied to those types of container which are subjected to sudden temperature changes during washing, pasteurisation or hot filling. The test method and equipment are described in ASTM Standard C149–77. The bottles, contained in a metal basket, are completely filled with water in a hot tank. After five minutes to reach thermal equilibrium, the basket containing the bottles is automatically transferred to a similar tank containing cold water. After 30 seconds the basket is raised and the number of fractures is noted. The ASTM Standard recommends a temperature difference of 42°C between the two tanks. Lomax reviews the relevant voluntary standards.

Two other strength measurements are carried out as required. Neither is defined by a standard but in most companies the same equipment is used. The more widely used measurement is an impact test. The container is held vertically against a backstop and a pendulum-like arm is allowed to swing so that its head strikes the side wall of the container at a point opposite the backstop. Lomax considers that the test is particularly valuable as part of strength verification procedures when a new container design is being developed and also for testing containers made by the press-blow process. As pointed out earlier, the internal surfaces of containers made by this method can be greatly weakened by foreign particles transferred to the glass from the plunger. Impact loading produces tensile stresses over relatively large areas on the internal surface. Hence the test is particularly suitable for determining whether an unacceptable amount of transfer has taken place. It should be repeated several times, the container being rotated between each test to ensure that most of the internal surface has been stressed.

Containers are subjected to vertical loads during filling and some bottlers will require a test for vertical load strength to be part of quality control. There are no national standards, but the equipment is simple. The British Glass Manufacturers Corporation suggest minimum values for vertical load strength of 6000 N for refillable and 4000 N for non-refillable containers.

An abnormally low result in any strength test will start a search for the cause – a surface defect, an inclusion or stone in the glass, poor annealing or cordy (inhomogeneous glass). Observations of the fracture pattern and of fracture surface markings (see Chapter 3) will help to locate the fracture origin and focus attention on surface defects in that region.

Examination of glass quality and determination of stresses due to inhomogeneity is carried out using a polarising microscope on a ring section cut from the glass. Stones (crystalline inclusions) in the glass may be identified using a similar microscope although it may be quicker to use X-ray powder diffraction. The state of annealing can quickly be checked on-line using a simple polariscope.

Wall thickness distribution is usually determined on a magnified image of a section through the container, the section being cut with a slitting wheel along the length or across the width of the container. Similar projection magnifiers are used to check dimensions, particularly those which are critical around the finish.

A considerable amount of design work is done, which interacts with the quality control activity. Initially the appearance of a new container is discussed between the manufacturer and the customer and considerable use is made of computer aided design to display and modify possible designs. Experience may suggest that some designs will be more difficult to manufacture than others or may give rise to strength problems. Now that fairly compact finite element programs are available, it is possible to predict the strength of new designs before manufacturing commences (Müller–Simon and Barklage–Hilgefort 1988).

Deterioration of the container contents is something that must be guarded against. Certain drugs in solution and blood plasma can be affected by alkali extracted from the glass. Photochemical reactions also affect some drugs and cause changes in the taste of drinks like beer and cider. BS 1679. Part 6: (1984) for medicine bottles specifies the maximum amount of alkali that may be extracted from the internal surface in a standard test and also requires that amber glass used for this purpose shall transmit no more than 10% of the incident radiation at any wavelength between 300 and 400 nm for a 2mm glass thickness. For beer and cider bottles, alkali extraction is not a problem but BS 6118: 1981 requires that the amber glass used should have a transmission of 30–50% at 550 nm for a 2 mm thickness.

Alkali extraction from the internal surface of containers can be greatly reduced by sulphating, a process in which the surface is exposed to an acid gas shortly after the container is released from the machine. This can be done very simply by dropping a pellet of ammonium sulphate inside each container. Other solids which decompose to produce acid gases are also effective (Persson 1962).

5.9. Competition from alternative materials

During the past twenty years, the glass container industry has lost some of its share of the market to other container materials, especially to plastic containers and metal cans (La Course, Varshneya and Alderson 1985). This is obvious from even a casual examination of supermarket shelves. The fall in demand for glass containers has not, however, been as great as might have been expected, since there has been a general increase in demand for containers of all kinds as supermarket shopping has developed. Currently the products of the UK packaging industry have a market value of £6.3 Bn, having grown by 60% over the past six years. During that period the market value for glass containers has remained almost constant at £415 M. It now accounts for about 6.5% of packaging products (Anon 1988b).

The polymer polyethylene terephthalate (PET) is a relatively new container material. It has an attractive clear appearance – very like glass in this respect – but initially it was not well received by the consumer. Diffusion of gases through the container wall was excessive and carbonated drinks and beers soon went flat. Modern PET containers are much better in this respect. PET has one obvious advantage over glass. A PET container of a given capacity is less than half the weight of its glass equivalent, though if one compares the weights of similar filled containers, the percentage difference is not so great.

Fracture of glass containers in the supermarket, though rare, presents problems which are difficult to accept and which do not arise with PET. The glass fragments scatter widely and much time is spent making sure that they have all been recovered. Accidents of this kind have also been known to cause serious injury, especially to children.

In spite of the disadvantages of glass as a container material, it is still widely used for many products and not only for high cost products such as wines and spirits. A recent survey indicates that there is a strong consumer preference for glass containers. However the industry recognises that it cannot rely on that preference to maintain its position. There must be progress towards solving the weight/strength problem if the packaging industry and the retail trade is to continue to accept the material.

5.1. Recycling and legislation

For a number of years the glass container industry has successfully projected an image of itself as being environmentally friendly and as being concerned with the conservation of materials and energy. This has been done mainly through its various schemes for recycling.

Waste glass has always been recycled within the factory. The material mixture (batch) fed into the furnace contains scrap glass (cullet) in the form of rejects originating within the factory. Typically this may amount to about 25% of the batch – more in some cases. Apart from making use of material that would have to be disposed of, the addition of cullet accelerates melting of the glass and saves a certain amount of energy.

Glass brought in from outside the plant (foreign cullet) can be used in the same way, provided that it is free from foreign material, especially metal, and is periodically analysed, so that any necessary changes can be made to the raw material mix to maintain a constant composition of glass being melted. There is no great problem in increasing the percentage of cullet in the batch to 50% – even 80% in some furnaces. Thus there is adequate capacity in the glass furnaces distributed throughout the country to accept a considerable amount of foreign cullet.

The major problem is not one of glass melting technology. It is to establish effective and economic arrangements to collect and transport the cullet from the individual home to the glass works. This has been done in two ways (a) by setting up a network of collection points for the containers (bottle banks) and (b) by processing domestic waste to extract the glass. In the UK, both systems are in use. Since a variety of individuals and public authorities are involved in handling the cullet, an important part of planning the system is to ensure that adequate incentives are provided along the chain to ensure that it is effective.

It is interesting that, but difficult to understand why, the percentage of glass recycled varies so much from one country to another. In 1976, the quantity recycled as a percentage of the total output of non- returnable containers was 56% in Germany, 62% in Belgium, 26% in France and 83% in Holland (Vermylen 1979). The author does not give the figure for the UK but it is believed to be less than 20%. There is no suggestion that the British are more lazy or less public spirited than other nations. A major factor contributing to the difference is that the disposal of waste in land-fill sites in the UK is far cheaper than in other countries (Cook 1986).

However effective these schemes may be in their respective countries, it seems very likely that they contribute towards giving glass a favourable image and thereby help the glass container industry to maintain its market share.

This may become even more important in the future as environmental issues come into greater prominence in national politics.

There are signs of how important this may become. A few European countries are already beginning to legislate against some types of container which cause environmental problems. In 1978 the Danish government banned the sale of carbonated beverages in non-returnable containers and announced their intention of banning the sale of canned beer. Such measures provoke strong reactions from the sectors of the container industry concerned, but it is clear that political decisions may have profound effects on the whole container industry, including the glass container industry. At present there is no indication that the UK government will develop its recent interest in environmental issues in this direction.

Chapter 6

Glass Seals in Vacuum- and Gas Discharge Devices

6.1. Introduction

Two industries have been of great importance in stimulating the development of new glasses – the electrical and the optical industries. In recent years their respective fields of interest have begun to overlap with the development of optoelectronics, and in this new technology, glass continues to play a major role.

Although glass–metal seals were used early in the nineteenth century in making scientific equipment, the first technological use of metal leads sealed through a glass vacuum envelope was by Edison in the USA and by Swan in England to make the first incandescent filament lamps. The success of the lamp development called for power generation and distribution systems large enough to light cities. The availability to physicists of glass vessels with sealed – in metal leads led to the classic experiments that determined the properties of the electron. So glass–to–metal seals have been important in various ways for the development of electronics and electrical engineering.

The development during the first half of the present century of new types of lamps and valves for both domestic and industrial applications called for a considerable number of new glasses. Although the introduction of semiconductor devices led to the replacement of some glass envelope devices, new applications for glass in the industry have arisen in their place (Francel 1986; Tummala and Shaw 1986).

A number of large glass companies still have a major involvement in supplying glass components to the electronics and lamp industries. The best known are Corning in the USA, Schott Glaswerke in West Germany and GB Glass Ltd in the UK. The technology of sealing glass to other materials is still an important one, which is relevant to a wide number of applications of the material.

Detailed accounts of all aspects of glass sealing are given in texts by Partridge (1949); Kohl (1967) and Espe (1968). The book by Cayless and Marsden (1983) contains useful information on the use of glass and on manufacturing processes in the lamp industry.

There are a number of properties of glasses (especially the optical properties) which are important for their application in the electrical and electronics industries. They are discussed elsewhere in this book.

6.2. Sealing glasses and metals

Compositions of the more important sealing glasses, with some indication of their properties, are given in Table 2. More complete information can be found in the general references given above and in the manufacturers' brochures. In addition to the multi-component glasses, silica glass is now widely used especially in the lamp industry in applications where high envelope temperatures are encountered e.g. in arc lamps for studio lighting and in tungsten–halogen lamps.

6.2.1. Properties of sealing metals and alloys

For sealing applications the most important physical property is the thermal expansion coefficient. Table 13 gives values for the more important sealing metals and alloys and Table 14 values for the most commonly used sealing glasses.

It is difficult to give here any more than a brief account of the circumstances under which each of these sealing metals, and their associated glasses, are used. The following notes may be helpful.

Tungsten and molybdenum have high electrical conductivities and are used in wire or rod form in conjunction with expansion-matching borosilicate glasses when leads capable of carrying high currents are required. In contrast to glasses, the metals have expansion coefficients that do not change very much with temperature. Consequently the stresses in seals made with these metals change considerably with temperature because of the difference between the shapes of the expansion curves of the two materials.

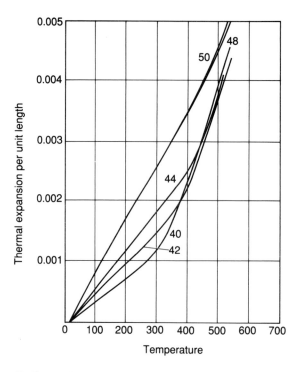

63. Expansion curves of some nickel-iron alloys. The number an each curve is the percentage of nickel *(Partridge 1949)*

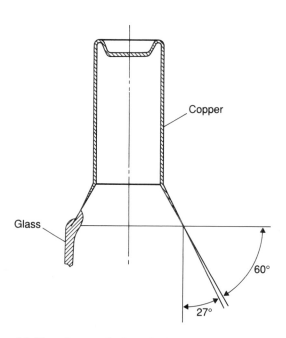

64. Housekeeper design of tubular seal *(Partridge 1949)*

In contrast the nickel alloys, especially the Ni–Co–Fe and Ni–Cr–Fe alloys, have expansion curves that are very similar in shape to that typical of a glass, the increase in slope occurring at the Curie temperature (Fig. 63).

Seals of these materials to the appropriate glasses can have very low stresses at all temperatures from the annealing temperature of the glass down to room temperature. The alloys are available in a variety of forms and are widely used to make a range of designs and sizes of seal, from minute terminal seals for low current devices to large tubular butt seals several centimetres in diameter.

Table 13
Expansion coefficients of sealing metals

	Expansion Coefficient $\times 10^{-7}\,°C^{-1}$	Temperature range °C
Tungsten	44	Room temperature
Molybdenum	51	Room temperature
Copper	165	Room temperature
42% Ni–Fe	46	25–100
46% Ni Fe	70	25–100
52% Ni Fe	99	25–100
Kovar, Nilo K etc. (ca 28% Ni,17% Co, rest Fe)	58	25–100
Nilo 475 (ca 47% Ni, 5% Cr, rest Fe)	98	20-100
Dumet (Cu coated 42% Ni-Fe)	92 (radial) 65 (axial)	

Table 14
Expansion coefficients of sealing glasses

	Expansion Coefficient $\times 10^{-7}\,°C^{-1}$	Temperature range °C
Fluorescent lamp tubing	92	50–300
Incandescent lamp bulbs	95	50–300
Dumet sealing lead glass	90	50–300
Kovar sealing	47	50–300
Tungsten sealing	37	50–300
Silica glass	5	50–300
Solder glasses (lead–zinc–borate)	76–88	50–200

Copper has too high an expansion coefficient to make matched seals of conventional design to any of the standard sealing glasses. However it is widely used in the Housekeeper seal design (Fig. 64) to seal to borosilicate glasses with expansion coefficients as low as 40×10^{-7} °C. This can be done, in spite of the large expansion mismatch, because the edge of the copper is thinned to a feather edge and ductile yielding of the metal prevents excessive stresses being set up in the glass. Oxygen free, high conductivity copper is used, otherwise the metal may become brittle as a result of heating in the flame.

The idea of using very thin metal to avoid high stresses in the glass is also applied in the molybdenum foil seal to make leads through silica glass lamp envelopes. The expansion mismatch in this combination is also very large. The foil is approximately 0.015 mm thick and its edges are feathered. Small single foil seals are used in lamps for car headlamps and home projectors. Large multiple seals of complicated design are used in high power arc lamps.

'Dumet' is universally used as the sealing metal in incandescent filament lamps. It is a composite material consisting of a core of a 42% or 43% Ni–Fe alloy with an outer sheath of copper, which amounts to about 30% of the weight of the wire. Because of the copper sheath, it has sufficient current carrying capacity for its application, but because of its composite structure, the thermal expansion properties are non-isotropic. It is well matched in the radial direction to the lead glass with which it is used, but the match in the axial direction is very poor. It can be used only in small diameters – less than 0.8 mm.

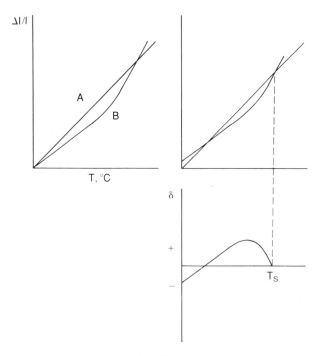

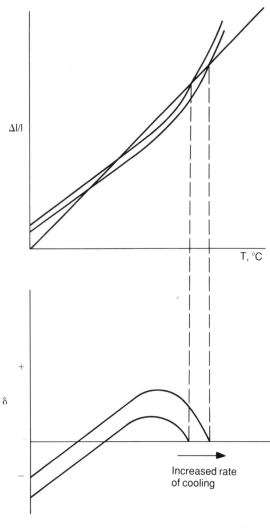

65. The dependence of the differential free contraction
(and hence the seal stresses) and the shapes
of the expansion curves of the seal materials (schematic)

66. Effect of cooling rate on the setting point and hence
on the stress-temperature curve (schematic)

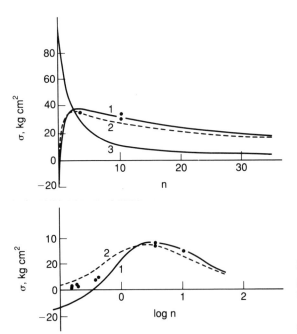

67. Dependence of the room temperature stress in a
glass-alumina sandwich seal on the stiffness ratio, n.
Curve 3 is calculated ignoring relaxation effects. Curves
1 and 2 are calculated including the effect of relaxation.
Circles are experimental results
(Rekhson 1979a)

6.2.2. Pre-treatment of sealing alloys

In order to make reasonably strong, leak-free seals, it is necessary to develop on the metal surface a dense, adherent oxide layer which will partly dissolve in the glass while the seal is being made. The metal surface should also be clean and free from metal-working defects which could act as leakage paths. Metal preparation is described in the references given in the introduction, but again it may be helpful to give a brief indication of what is involved.

Copper and Dumet are probably the most troublesome materials. When heated in air they develop a poorly adhering, black layer of cupric oxide which is completely unsatisfactory as a seal interface. They are therefore 'borated', i.e. dipped into an aqueous solution of borax and then heated to give a protective film of sodium borate glass. The film limits the access of oxygen to the metal surface during sealing. The correctly oxidised surface consists mainly of cuprous oxide and is bright red or orange in colour.

Decarburisation and controlled pre-oxidation of the nickel alloys is usually recommended, a typical schedule involving firing in wet hydrogen for 30 min at 1000°C. A number of research studies have been made of this process (Notis 1962, Abendroth 1965, Piscatelli et al. 1976). Many bubbles at the seal interface are an indication that the metal has been insufficiently decarburised. The 'correct' seal colour is medium grey, except for the chromium containing alloy where the interface shows the green colour of chromic oxide. This oxide gives excellent glass–metal adhesion at the price of difficulties in subsequently making electrical connections to the exposed wire.

Tungsten and molybdenum are cleaned chemically or electrolytically prior to sealing. It is easy to over-oxidise these metals , especially molybdenum, during sealing. When strongly heated in air, molybdenum emits copious fumes of the trioxide and this has to be prevented by restricting the access of air, especially when making seals to silica. When sealed to borosilicate glasses, the seal colours are golden brown for tungsten and chocolate brown for molybdenum. The foil seals of molybdenum to silica appear to be free from oxide at the interface. They have the bright, shiny appearance of unoxidised metal. It is possible, but not certain, that the interface composition is a molybdenum silicide.

6.3. Seal stresses (Varshneya 1982)

In all seals the thermal expansion curves of the two materials differ and consequently there will be stresses in the seal over a range of temperatures. The possibility of fracture of the glass has to be guarded against by ensuring that the expansion curves of the two materials are reasonably well matched and by paying attention to the seal design. It is not possible to specify an upper limit for stress in the glass which is generally applicable but, in most applications, the seal will be satisfactory if the maximum tensile stress throughout the operating temperature range is less than 10 MPa.

Fig. 65 shows how the stresses in a typical seal may vary with temperature. Most metals and ceramics have an approximately linear expansion curve (A) whereas a glass shows a change in slope in the transformation range (B). The materials are sealed together at a temperature which is usually well above the transformation range, when the viscosity of the glass is too low for any stress to be set up. Stresses arise only when the glass is cooled to below the transformation range.

A simple geometrical construction gives an approximate indication of how the stresses will vary with temperature. Curve B is displaced vertically to intersect curve A at the setting point. It is assumed for simplicity that this is a well-defined temperature, above which the seal is stress free and below which there is no stress release in the glass due to viscous flow or any other mechanism. This is an oversimplification. As is shown below, the 'effective setting point' depends considerably on the rate at which the glass is cooled through the transformation range.

The vertical distance between the expansion curves in Fig. 65, δ, is termed the differential free contraction. It is the difference in length per unit length of the materials at a specified temperature assuming that they have the same length at the setting temperature and that they have been free to contract independently. At any point in a seal and at temperatures below the setting point, the stress component in any direction is directly proportional to δ.

Fig. 66 shows the dependence of the setting point on the rate of cooling. This also affects the shape of the expansion curve of the glass. Thus the geometrical construction of Fig. 65 is of only limited value for determining the shape of the stress–temperature curve. The marked dependence of the stress–temperature curve on the rate of cooling and on heat treatment has been studied experimentally in detail by Oldfield (1959) for seals made from borosilicate glasses.

A number of papers show how the effects of heat treatment on seal stresses can be accounted for by using information on the viscoelastic properties of the glass (Rekhson and Mazurin 1977; Rekhson 1979a, 1979b; Scherer and Rekhson 1985). This work also has an important bearing on the effect of seal dimensions on seal stresses, an aspect which had previously been believed to be entirely explainable in terms of linear elasticity.

This is illustrated in Fig. 67 for glass–ceramic sandwich seals having various values of the stiffness ratio, n (the ratio of glass to ceramic thickness). Curve 3 is calculated assuming that δ is independent of n. Curves 1 and 2 are calculated with slightly different theoretical models both incorporating the effect of ceramic stiffness on the stress relaxation. Clearly the effect is a significant one and can be of considerable practical importance.

6.4. Seal stresses and seal design

Many designs of seal are in use. In general the stress distribution in the glass is three dimensional, each stress component varying in magnitude and direction throughout the glass.

Until recently, the analysis of seal stresses was limited to simple geometries, e.g. the axially symmetric design in which a metal wire is sealed inside a glass tube (the bead seal). The solution to simple stress analysis problems of this kind (or very similar ones) can be found in standard text books on elasticity theory. Texts dealing with thermal stresses are particularly helpful since there is no essential difference between a problem in which there is a sudden temperature discontinuity in a homogeneous elastic solid and one in which the solid is at a uniform temperature, but is made from materials of different expansion coefficients.

Theories giving the stress distributions in a number of simple seal designs have been developed and results are available in the form of design curves and/or equations for the variation of the stresses with position in the glass (e.g. Poritski 1934; Hull and Burger – bead seal 1934; Hull 1946 ; Adam 1954 – window seal; Rawson 1949 – concentric tube seal; Rawson 1951 – butt seal).

Fig. 68 shows the design curves for the bead seal. They show the effect of the radius ratio b/a on the stresses in the glass at the interface for unit positive value of δ. Note that whilst the circumferential and axial stresses are compressive, the radial stress is tensile.

The same theory, but with an appropriate exchange of elastic constants, is used to calculate the stresses in a window seal, in which a cylinder of glass is sealed inside a tube of metal. It is an important fact that in this design, for positive δ, all three stress components are compressive and are constant throughout the glass. This is a very strong design of seal which has been widely used in observation windows for pressure vessels. In a variant of the design, but still using the outer ring to put the glass into compression, current carrying leads are sealed through the glass, as shown in Fig. 69. The leads are made from an alloy which closely matches the glass in expansion so that the state of uniform compression in the glass, characteristic of the simple design, is only slightly disturbed. This construction is also extremely strong and has been used in such demanding applications as leading current through the pressure vessels of nuclear reactors. It is necessary when using compression–seal designs to ensure that the yield stress of the metal of the outer ring is not exceeded otherwise the stress distribution in the glass will not be as intended and failure is possible.

Provided that the simple theories are used with an awareness of their limitations, they are still of some value. However, as we have seen in the previous section, one cannot expect to be able to predict the seal stresses accurately without taking into account the effects associated with relaxation phenomena in the glass. Also the simple theories do not give the stresses near the ends of the seal, which are usually significantly higher than those near the centre.

Practical workers with glass seals know how important the end stresses are and take care to ensure that glass in this region is shaped so as to avoid re-entrant angles and to encourage any crack starting at the end to run out of the side of the seal rather than propagating along the length. Fig. 70 shows examples of good and poor seal shapes for the case of a bead seal.

Now that numerical methods of stress calculation based on the finite element method are readily available, there is no longer any difficulty in calculating end stresses. The methodology for seal stress calculation now appears to be fully developed, provided that the relevant glass property data is available (Varshneya and Petti 1978; Gulati and Hagy 1978a,b).

6.5. The control of sealing materials
(Anon 1949; ASTM Designations F144-73; F 14-68; F 140-73; Hagy and Smith 1969; Hagy 1979)

Control by direct measurement of the expansion properties of sealing metals and glasses is a relatively insensitive method of ensuring that seals made from these materials will have acceptably low stresses. A far better method, and one which is widely used, is to make a seal of standard dimensions and to measure, by a photoelastic method, the stress at a chosen point on the seal interface as it cools at a controlled rate from above the setting point to room temperature.

Various standard designs of seal are used: the sandwich seal (Fig. 71a) used if the metal is supplied as sheet and the bead seal (Fig. 71b) if it is supplied as wire. A homogeneous stock of the corresponding sealing glass is kept as a reference standard to make the seals. To control a sealing glass, rather than a sealing metal, the block seal (Fig. 71c) or trident seal (Fig. 71d) may be used. The last two are glass–to–glass seals in which one component is from a stock of standard glass, the other being the glass under test.

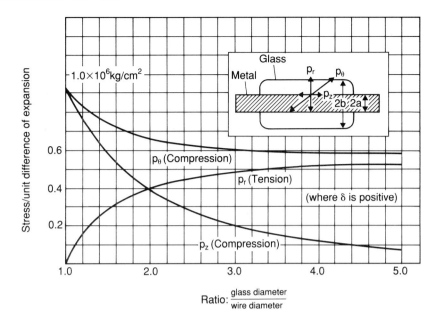

68. Relationship between the interface stresses in the central region of a bead seal on the ratio of the glass to wire diameter
(Hull and Burger 1934)

69. Large compression seals for nuclear reactor applications
(Schott Glaswerke)

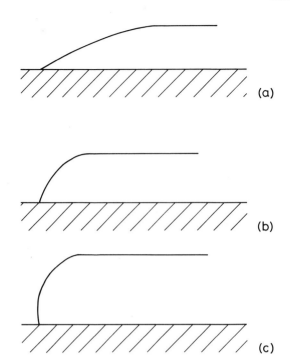

(a)

(b)

(c)

70. Examples of good (c) and poor (a & b) seal shapes

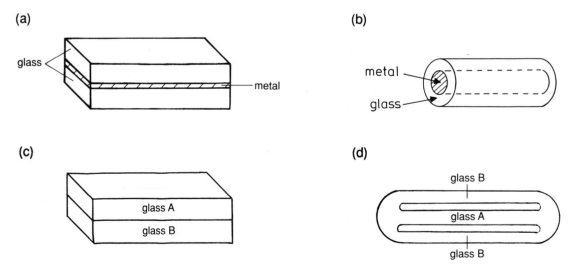

(a)

glass

metal

(b)

metal

glass

(c)

glass A

glass B

(d)

glass B

glass A

glass B

71. Standard seal designs.
(a): sandwich seal (b): bead seal (c): block seal
(d) trident seal

6.6. Sealing techniques

6.6.1 Solder glasses (Forbes 1967; Frieser 1975; Takamori 1979; Rabinovich 1979)

In many types of device, and in the lamp industry in particular, seals are made by flame fusion. However there are applications where the use of flames is to be avoided because of the need to prevent deformation of the glass components or damage to components inside the device. This need first arose in the manufacture of miniature radio valves during World War II. The appropriately named 'solder glasses' were developed in a number of countries (Dale and Stanworth 1948, 1949). They have compositions based on the $PbO-ZnO-B_2O_3$ system, are a good expansion match to standard soda- lime glasses but have a much lower viscosity so that they flow out and can make joints at temperatures below 600°C.

They are still widely used, but later 'devitrifiable' solder glasses were developed, based on the same system, mainly for use in sealing the face plate to the cone of colour TV tubes. After making the seal, the tube is given a heat treatment to cause the solder glass to devitrify in a controlled way. It can then be heated to a somewhat higher temperature for bake-out than would otherwise be possible.

Fig. 72 (p. 78) shows a flat panel plasma display tube in which a solder glass is used to assemble the components. Problems of alignment and registration are quite severe in making such a device and one can readily appreciate from this example the advantage that the use of solder glass sealing can offer.

The glasses are easily melted at comparatively low temperatures (below 1000°C) and, after pouring into water, the material is milled to a powder. The glass is not so resistant to atmospheric attack as other commercial glasses and one sometimes finds that the flow properties deteriorate if powdered solder glass is stored for several months. It is preferable to mill batches of glass only as required and to store the powder in air-tight jars. The powder is applied to the surfaces to be joined as a dispersion in a liquid binder though for some applications, the powder may be pressed into a preform.

6.6.2. Furnace seals

Seals of the types shown in Fig. 73 (p. 78) are widely used to encapsulate semiconductor devices and small quartz crystal oscillators. The metal components and the glass are assembled in graphite moulds and passed through a furnace, typically at 900°C, in which there is a protective atmosphere to prevent excessive oxidation of the sealing metal or the graphite moulds. Graphite is an excellent material for this application since it is easy to machine and is not wet by molten glass. The glass may be in the form of cut lengths of rod and tubing of controlled weight, but it is now more common, especially for the smaller seals, to use pressed and sintered preforms made from the powdered sealing glass. Some companies manufacture the complete seal to customers' designs whilst others supply only the preforms. Sintered seals are not transparent and cannot therefore be inspected for stress by the usual photoelastic methods. However this difficulty can be overcome if necessary by making control seals using the same unmilled sealing glass. The much larger seals in Fig. 69 were also made by the furnace method.

6.6.3. Infrared sealing (Hoogendorn and Sunners 1969)

Fig. 74 (p. 78) shows an alternative to the solder glass method, suitable for sealing very small devices in which the internal components must be protected from damage. Reed switches used in telephone exchanges are made using a glass which strongly absorbs infrared radiation. This is achieved by adding one or two percent of iron oxide to the glass and then melting under reducing conditions to ensure that the iron is present in the IR- absorbing ferrous form (Lian et al. 1982). A protective atmosphere prevents oxidation of the switch electrodes. The heat for sealing is supplied by a tungsten halogen lamp, the radiation from which is focused onto the seal.

6.6.4. Field assisted sealing (Wallis 1974, Younger 1980)

This novel technique has the capability of producing strong vacuum tight glass–to–metal seals at temperatures significantly below the annealing temperature of the glass. Thus the risk of deformation of the glass is completely eliminated. The faces to be sealed together are first lapped flat and smooth. They are placed in contact and a d.c. voltage of about 1 kV is applied across the interface, the glass being the negative electrode. After a few seconds, a strong hermetic bond is formed. The mechanical force produced by the electric field pulls the surfaces into closer contact and some electrolysis occurs causing transport of material across the interface. It is claimed that the technique works well with a range of sealing glasses and that it may also be used to bond glass to glass.

Sealing techniques involving glass–ceramic materials are described in Chapter 8.

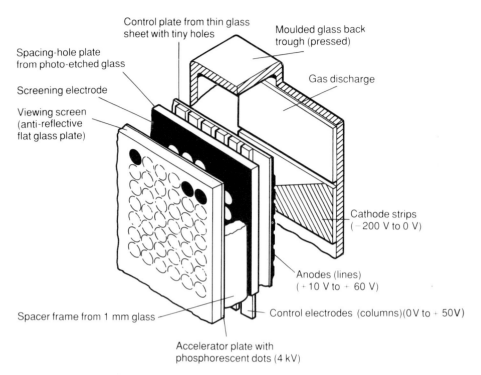

Spacing-hole plate from photo-etched glass

Control plate from thin glass sheet with tiny holes

Moulded glass back trough (pressed)

Screening electrode

Gas discharge

Viewing screen (anti-reflective flat glass plate)

Cathode strips (−200 V to 0 V)

Anodes (lines) (+10 V to + 60 V)

Spacer frame from 1 mm glass

Control electrodes (columns)(0 V to + 50V)

Accelerator plate with phosphorescent dots (4 kV)

72. Elements of a flat panel monitor
(Schauer 1982)

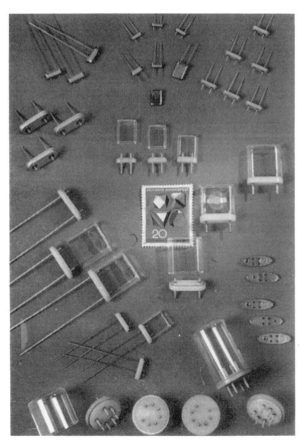

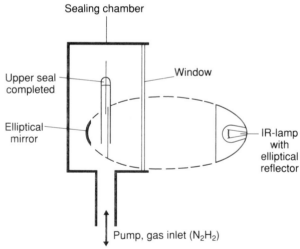

Sealing chamber

Upper seal completed

Window

Elliptical mirror

IR-lamp with elliptical reflector

Pump, gas inlet (N_2H_2)

74. Sealing chamber for making reed switches by focused infrared radiation
(Paschke 1978)

73. Glass encapsulations made using sintered glass bases
(Schott Glaswerke)

Chapter 7

Coatings on Glass

7.1. Introduction

Many kinds of coating, applied by various techniques, are of importance for modifying the properties, enhancing the performance and increasing the range of applications of glass. Some have already been mentioned in the chapters on flat glass and containers.

The subject of coatings on glass is so large and varied that it is necessary to limit what will be covered here. No account will be given of coatings applied by vacuum evaporation to produce anti-reflection coatings and multi-layer interference filters. Coatings of this kind are well known and their technology is well developed. For a more complete treatment, covering all kinds of coating, the reader is referred to the excellent book by Pulker (1984).

7.2. The chemical process for mirror manufacture (Schweig 1973)

The process currently used for the manufacture of mirrors is based on that invented by von Liebig in the mid nineteenth century. Fig. 75 shows the multi-layer structure of a typical commercial coating. (Thomas, Pitts and Czanderna 1983).

The glass sheets are fed along a horizontal conveyor on which they are subjected to a series of treatments:

1. Cleaning: the glass is scrubbed with brushes fed with a suspension of cerium oxide in demineralised water.
2. Sensitisation: the glass is sprayed with an acidified solution of $SnCl_2 \cdot 2H_2O$. Without this step, the subsequent silvering process would be ineffective.
3. Silvering: the glass is sprayed simultaneously with two solutions. That containing the silver is an ammoniacal solution of silver nitrate. The second acts as a reducing agent and is a solution of an invert sugar or glucose.
4. Coating with copper: this layer provides some protection of the silver layer in the finished mirror against chemical attack by traces of sulphur and chlorine compounds in the atmosphere. A suspension of zinc or iron powder in a copper sulphate solution is sprayed onto the glass surface.
5. Backing paint layer: after the glass has been dried, it is passed through a curtain coater, followed by an oven, to apply a backing layer of polymer. It is common to apply two such layers.

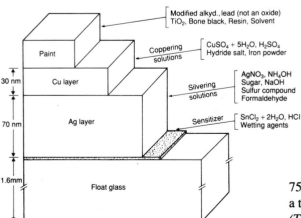

75. Layer structure and chemical components used in a typical commercial mirror
(Thomas et al. 1983)

The process is relatively easy to operate, given good control of the spray solutions and clean operating conditions. After each of the wet stages, it is essential to flush the glass thoroughly with demineralised water. Dusty conditions are to be avoided.

The mirror coating must be applied to the air-contact surface of the float glass. If one attempts to coat the tin-contact surface, the results are unsatisfactory. This is surprising. One might expect that the significant concentration of tin present at the surface of the tin-contact surface of float glass would be sufficient to provide the sensitising action necessary to deposit the silver layer. This is not so. Clearly the state in which the tin is present at the glass surface is important – an aspect which has been studied in detail by Pederson (1982).

The main use for chemical mirrors is in the home, where the durability requirements are not particularly severe, except perhaps in the bathroom where condensed moisture films may persist on the rear, coated surface. The products have to meet standard specifications, which include optical tests for flatness and accelerated chemical corrosion tests involving temperature cycling in a salt-spray cabinet (e.g. BS 5466 Parts 1 or 3 1977)). Safety is ensured by applying a tough polymer backing layer. The finished mirror is required to meet a similar strength and fracture specification to that which applies to other forms of safety glass used in buildings.

In a recent development, mirrors made in the way described are expected to withstand far more stringent operating conditions. Several solar power stations are being constructed, mainly in the USA but at least one in France. They use commercial glass mirrors to reflect the sun's radiation onto a boiler mounted on a central tower. The installations are quite large. For example that at Barstow, California, will have a mirror area of 93,000 m^2 and an output of 10 MW. At an early stage of the project it was discovered that the mirrors were deteriorating and there was no possibility that they would reach their intended service life of 30 years.

A very large programme of research was mounted to identify the mechanisms of attack and to study in detail the chemistry of each stage in the mirror-making process. The work has involved the application of all the modern methods of surface analysis which are available. Probably no other problem in the field of the surface science of glass has been studied so thoroughly. Those results which are the most interesting for the insight they give into surface chemistry and the process chemistry of mirror manufacture are to be found in papers by Pederson and Thomas (1980); Shelby et al. (1980); Pederson (1982) and Thomas et al. (1983)

7.3. Base exchange processes

In a base exchange process, the alkali ions in the surface of the glass are replaced by other ions from a source which is in contact with it. The source material may be a fused salt or a carrier paste containing the ion to be introduced. The temperatures required are moderate, often below the annealing temperature of the glass. In certain situations it is possible to accelerate the process and produce a thicker exchanged layer by applying an electric field. Applications include glass strengthening (containers, windscreens, domestic glassware, safety lenses), surface colouring (staining) and, most recently, gradient index optics and certain types of optical waveguide.

Fundamentals of the process, including both the kinetic and the thermodynamic aspects, are dealt with in several reviews (Garfinkel 1968; Doremus 1969, 1973; Terai and Hayami 1975; Beier and Frischat 1985; Ramaswamy and Srivastava 1988).

Weyl (1951) gives a brief explanation of the factors that affect the extent and the rate of exchange. It is assumed that the glass is heated in contact with the compound RX where R^+ and X^- are univalent. Base exchange between Na^+ ions in the glass surface and the R^+ ions involves the reaction:

$$Na^+ \text{ (silicate glass)}^- + RX = R^+ \text{ (silicate glass)}^- + NaX$$

During reaction an equilibrium is reached at the glass surface as a result of which the following relationship exists:

$$\frac{C(R^+ \text{ in glass})}{C(Na^+ \text{ in glass})} = K. \frac{\text{Activity of } R^+ \text{ in coating}}{\text{Activity of } Na^+ \text{ in coating}}$$

The value of K varies with temperature and is determined by thermodynamic factors.

The rate of diffusion of the R^+ ions into the glass is determined by the interdiffusion coefficient, D, which is temperature dependent. D is given by:

$$1/D = N_A/D_A + N_B/D_B$$

where the N and D values are respectively the fraction and the diffusion coefficient of each ion A and B.

Weyl reviews the results of early experimental work, which are still of interest to anyone using the process. Thus silver exchange is far more extensive when RX is AgCl than when AgI is used. Also the contamination of the AgCl source melt by only 1%NaCl reduces the extent of base exchange to only a quarter of that achievable with a fresh AgCl melt. Both observations illustrate the importance of the thermodynamic factor.

7.3.1. Staining

Yellow or red colours can be produced by applying to the glass surface a paste containing respectively a silver or a copper compound and firing for about 30 min at 500°C. The chloride or sulphide are normally used, mixed with an inert carrier such as clay or ochre. The colour will be weak unless the glass contains an oxide which can act as a reducing agent, e.g. As_2O_3. Alternatively the colour may be intensified after the base exchange treatment by firing the glass in a mildly reducing atmosphere.

The colour is due to a colloidal dispersion of copper or silver particles, formed by reduction of the ions which have diffused into the glass surface (see Chapter 2). Staining is a very old process, introduced into Europe from North Africa in the ninth century and used during the middle ages to produce the yellow and amber colours in the windows of large churches and cathedrals in northern Europe.

Recent studies of the process include the work of Rawson (1965) and Meistring et al. (1976). Yates (1974) describes an interesting example of the use of field-assisted ion exchange to colour float glass. The process is carried out on the tin bath, which acts as one electrode. The second electrode is a copper bar mounted just above the glass ribbon. Contact between the upper electrode and the glass is through a molten layer of a copper–lead alloy. The process was used to make bronze-tinted solar control glass.

7.3.2. Ion exchange strengthening. (Bartholomew and Garfinkel 1980)

Ion exchange strengthening involves the replacement of R_1^+ ions in the glass surface by larger R_2^+ ions. This tends to expand the glass structure in the surface layer, a tendency which is resisted by the unchanged interior. As a result, a compressive stress is produced in the glass surface. The process must be carried out below the transformation range so that stress release in the glass during the process is relatively small. Fig. 53 compares the stress distributions through the glass thickness for thermally tempered glass and ion exchange strengthened glass. The surface compression layer in the latter is clearly far thinner. Optimising the heat treatment involves a compromise. A high temperature gives a deep exchanged layer, which is necessary for a high surface stress. On the other hand, at the higher temperatures, more stress release occurs during the treatment.

The process has the great advantage over thermal tempering that it can be used to strengthen much thinner glass. It has the disadvantage, however, of being very slow, treatment times being normally of the order of several hours. Much shorter times are used in the ion exchange strengthening of glass containers (Chapter 5) but the degree of strengthening so obtained is limited. About twenty years ago there was a major investment in the development of automobile windscreens using ion exchange strengthened glass (Blizard and Howitt, 1970), but this was made less attractive following developments in methods of thermal tempering of thin glass. There is still some use of the material in glazing for transport applications, but the most widespread application is for strengthening spectacle lenses for safety goggles (Stroud, 1988).

As might be expected, the glass composition has a large effect on the extent of the exchange which occurs and hence on the degree of strengthening that can be obtained. The highest strengths (700–1000 MPa) have been obtained using alkali aluminosilicate glasses containing 10–20% of alumina. Commercial soda–lime glasses and other alumina-free silicate glasses can also be strengthened but not to the same extent (Ohta 1975).

Early papers by Burggraaf et al. (1964, 1966); Spoor and Burggraaf (1966); Garfinkel (1969); and Garfinkel and King (1970) describe the relationship between the ion concentration and stress profiles near the glass surface for a range of glass compositions and treatment conditions. The effect of heat treatments after the base exchange on the surface stresses and hence on the strength was also studied. Zijlstra and Burggraaf (1968/1969) deal mainly with the strength and fracture behaviour.

Fig. 76 shows two Na^+ ion concentration profiles in a $Li_2O–Al_2O_3–SiO_2$ glass which had been base exchanged for 4 h at 415°C in a $NaNO_3$ fused salt bath. The concentration is expressed as a ratio relative to the surface concentration. The solid circles are results obtained after the glass had been heated for a further 4 h in air at 415°, the change being the result of further diffusion of the sodium ions into the glass, This reduction in the ion concentration gradient together with stress relaxation by viscous flow reduces the strength of the material at elevated temperatures. For this particular glass and this heat treatment, the measured concentration profiles lie very close to the theoretical curves, calculated assuming a constant interdiffusion coefficient. Such ideal behaviour is not always observed.

Burggraaf and Cornelissen (1964) showed that the stresses in the treated glass could be calculated from the concentration gradient by using a slightly modified form of the equations normally used for calculating stresses due

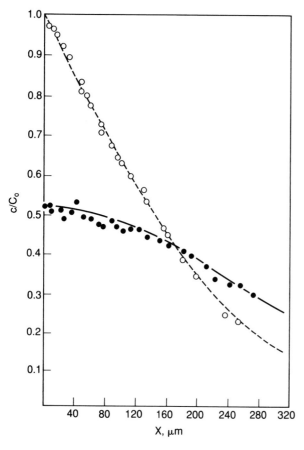

76. Concentration fraction as a function of distance from the surface.

○ Na exchange at 415°C for 4 h

● Na exchange at 415°C for 4 h followed by reheating in air at 415°C for 4 h. Curves are calculated from theoretical equations

(Garfinkel and King 1970)

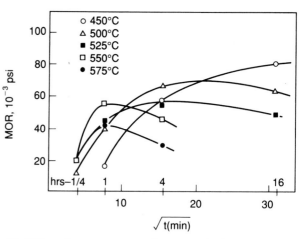

77. Effect of temperature on the strength of potassium-exchanged 1.1 $Na_2O . Al_2O_3 . 4 SiO_2$ glass
(Garfinkel 1969)

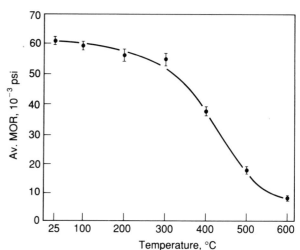

78. Abraded strength of sodium-exchanged 11 $Li_2O . 16.5 Al_2O_3 . 72.5 SiO_2$ glass after reheating in air at various temperatures
(Garfinkel 1969)

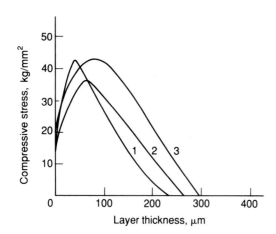

79. Relaxed stress profiles after ion exchange in $Na_2SO_4 + ZnSO_4$ (glass composition in mol%: 66.66 SiO_2, 16.66 Al_2O_3, 16.66 Li_2O)
1. 590°C 10 min; 2. 580°C 15 min; 3. 585°C 25 min.
(Zijlstra and Burggraaf 1969)

to temperature gradients, the modification being to replace the thermal expansion coefficient in the thermal stress equations by the dilation constant, a measure of the volume change produced per unit quanity of ion exchanged.

Spoor and Burggraaf (1966) studied the kinetics of stress relaxation after ion exchange and concluded that if the base exchange treatment temperature is less than 100°C below the strain point of the glass being treated, the build up of stress will be significantly reduced by relaxation. Fig. 77 from Garfinkel (1969) shows the effect of treatment temperature and time on the strength of a K^+ exchanged glass of molecular composition $1.1 Na_2O \cdot Al_2O_3 \cdot 4SiO_2$. Fig. 78 shows the decay of strength of a Na^+ exchanged lithium aluminosilicate glass on heating for 4 hr after treatment at temperatures up to 600°C.

More recent work on the process has been reviewed by Schaeffer (1985). He draws attention to the wide range of stress profiles which can be obtained by multistep and even single step treatments. Thus it is possible to produce a situation in which the stress rises to a maximum at some distance below the surface (Fig. 79). He suggests that such a stress profile may hinder the propagation of any crack moving into the material from the surface.

Yet another type of profile is obtained by field-assisted base exchange as shown in Fig. 80 (Abou-el-Leil and Cooper 1979; Götz et al. 1979).

7.3.3. Ion exchanged glass waveguides (Findakly 1985; Ramaswamy and Srivastava 1988)

In addition to the better known application of glass in optical communication fibres (Chapter 10), there is also considerable interest in developing compact optical devices for signal processing (Walker 1987; Venghaus 1989). The optical signal is transmitted from one part of the optical circuit to another through waveguides, one form of which consists of a strip of material within the glass surface, which has a higher refractive index than the surrounding material. The guide is made by electrically assisted diffusion from a salt bath source. When necessary, the lateral extent of the guide may be limited by applying an evaporated film metal mask before the treatment. One of the first papers in this field (French and Pearson 1970) is still of interest.

As for ion exchange strengthening, a great deal of research has been done to investigate the kinetics of the process for a number of glasses and diffusing ions and to study those factors which determine the refractive index profile within the treated region. The references given above summarize this work and also describe methods of manufacture and methods for measuring guide performance.

The ions producing the largest refractive index changes are Tl^+ (Δn ca 0.1) and Ag^+ (Δn ca 0.13). However Tl^+ is toxic and Ag^+ causes large signal losses if it is reduced in the glass to the metal. Exchange of one alkali ion by another gives much smaller values of Δn. Most work has been done on guides made by Ag^+/Na^+, K^+/Na^+ and Cs^+/K^+ exchange. By double exchange, i.e. following the first exchange with a second, it is possible to make 'buried' guides in which the refractive index passes through a maximum at some distance below the surface. Such a guide has better propagation characteristics and is less affected by surface irregularities in the glass. Accounts of recent work on Ag^+ ion exchange have been given by Fainaro et al. (1984) and Chludzinski et al. (1987).

Propagation losses down to 0.1 dB/cm have been obtained. Presumably these are considered low enough for the short transmission distances involved. They are very high compared with the losses of some of the multicomponent glasses made during research on materials for the communication fibre application.

7.4. Magnetron sputtering (Gläser 1980a, 1980b, 1983, 1989; Pulker 1984; Hartig et al. 1983)

7.4.1. Coating equipment

Much of the low-emissivity and solar control flat glass, described in Chapter 4, is made by this process. The equipment is capable of treating very large sheets, typically up to 6 m by 3 m in dimensions. The glass passes in steps through a series of treatment chambers, each of which can be isolated from its neighbours by vacuum locks so that the atmosphere in each zone can be controlled independently. The number of chambers required depends on the structure of the coating and the need of the manufacturer to produce a range of coating types. A widely used low-emissivity coating consists of a layer of silver sandwiched in between two layers of tin oxide. Thus, for making this coating, there must be at least three sputtering chambers, one for each layer of the coating (Fig. 81). The production rate may be as high as 1.86 million square metres per annum for a large 15 cathode system.

Until recently, the sputtering source was invariably a water cooled plate of the coating metal, mounted above and parallel to the glass at a distance of about 0.3 m. This is maintained at a negative potential of several hundred volts. When material is to be sputtered with no change in composition, the gas in the chamber is argon at a pressure of 10^{-3} to 10^{-2} mbar. Argon ions are produced in a glow discharge in the space between the target and the glass. These bombard the target, sputtering off atoms of the target material and electrons. The sputtered atoms strike the glass surface, where they become attached. The electrons emitted serve to produce more argon ions in the glow discharge and so maintain the cycle.

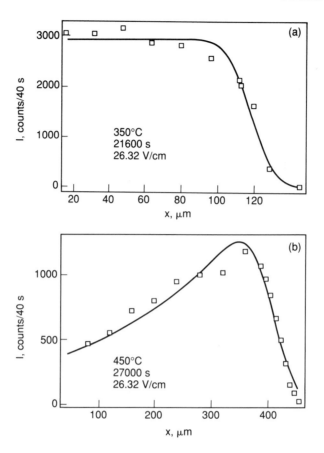

80. Diffusion profiles of silver in flat glass after electromigration. (a). 350°C 6 h; (b). 450°C 7.5 h. 26 v/cm
(Götz et al. 1979)

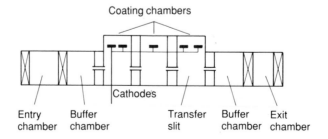

81. Semi-continuous vacuum sputtering plant
(Gläser 1983)

By placing an array of permanent magnets behind the source plate, the emitted electrons are trapped in a space close to the front of the plate. This results in a high ion density in this region and a high sputtering rate. Fig. 82 (Pulker 1984) shows a planar magnetron cathode in cross section. With this arrangement, sputtering of the target plate is non-uniform and a 'race track' shaped groove develops in the surface of the plate. Eventually the groove becomes so deep that the target must be replaced. Thus with this design, the utilisation of target material is low – in the range 20–30%. Using the newly introduced rotatable cathode, sputtering is more uniform and the utilisation figure is increased to 80%. Note that although the cathode rotates, the magnets are in a fixed position and so the cathode surface is moving continuously through a stationary plasma. Another advantage of the rotatable cathode is that it makes possible the sputtering of materials that cannot be sputtered using the planar design (McBride and de Bie 1991).

7.4.2. Some applications of sputter-coated glass

Low emissivity coatings

In recent years a range of sputter-coated flat glass products has become commercially available and the development of new products is continuing. In addition to the solar control and low emissivity products referred to in Chapter 4, modified versions of the conventional low emissivity coating are being introduced into car windshields. The silver film within the coating has a low sheet resistance of 4 to 8 ohms per square and 1500 watts can be dissipated over a 10 square feet area of windscreen, enough to melt approximately 0.1 in of ice in two minutes in sub-zero weather. Coated solar control glass is also being introduced into cars to improve comfort and reduce the load on air-conditioning equipment. It is necessary for this application that the coating should not reduce the visible light transmission below that required for safe driving (75% in Europe). Now that coatings are available which will withstand the temperatures involved in bending the glass after coating, further application of coated glass in cars is to be expected (Szczyrbowski *et al.* 1987). With the introduction of the rotatable cathode, it becomes easier to sputter materials such as SiO_2 and TiO_2. By applying layers of appropriate thicknesses of these oxides onto aluminium, high reflectance mirrors can be made. Until recently such components used in advanced technical equipment were made by batch processes using the electron-beam process. Thus there are many indications that sputtered coatings are being developed for use in a number of new markets.

The optical properties of coated glass depends on the optical constants of the coating materials and on their thicknesses. These constants vary with wavelength, especially for metallic materials. Figs. 83 and 84 respectively show the wavelength variation of the transmittance and reflectance of a low emissivity coating in the visible and near infrared. The reflectance continues to rise with increasing wavelength, reaching a value of 90% in the range 10 to 15 μm. The emissivity, e, at wavelengths where the glass is opaque (>5μm) is given by:

$$e = 1 - \rho$$

where ρ is the reflectivity. Thus for this material the normal emissivity is about 0.1. This compares with a value of 0.845 for uncoated glass. The determination of the emissivity of low emissivity glass is described in BS 6993 Part 1 and in a paper by Nicoletti, Geotti-Bianchini and Polato (1988). It involves the measurement of the reflectivity over the wavelength range 5 to 55 μm and the calculation of a mean whereby the measured values are weighted using corresponding values of the power distribution of a full radiator at a temperature of 283°K. This is too lengthy a procedure for routine process control. For this purpose a much simpler and quicker method appears to give satisfactory results. There is an almost linear relationship between the long wavelength emissivity of the coated glass and the film conductivity as measured by a commercial four-point probe instrument of the type normally used for measuring the conductivity of semiconductor silicon slices. In effect one is measuring the thickness of the silver layer which is the main factor determining the long wavelength emissivity.

Apart from the main requirement of low emissivity, it is also necessary to ensure that the colour of the coated glass is consistent. The material coated with the conventional $SnO_2/Ag/SnO_2$ layer appears slightly purple in reflected light and this reflection colour is sensitive to the relative thickness of the oxide and metal layers. If one could easily measure the thicknesses and had reliable values for the optical constants of the layer materials, one could calculate the optical properties of the coated glass as a function of wavelength and hence obtain a better understanding of the factors which determine those properties. A limited amount of work of this kind has been carried out (Fan *et al.* 1974; Rawson 1986, 1989). Fig. 85 shows the effects of the thicknesses of the silver and tin oxide layers on the reflectance in the visible and near infrared. These calculations were carried out using the optical constants determined for *evaporated* rather than sputtered silver films. The results probably give only a semi-quantitative indication of the effects of layer thicknesses; they cannot be expected to be exact.

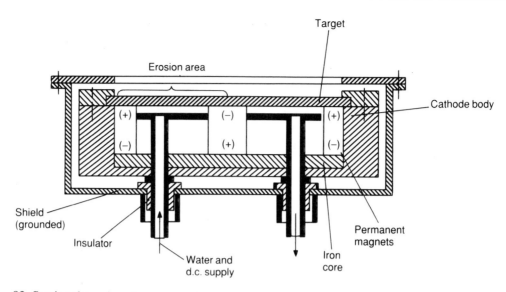

82. Section through a planar magnetron target
(Pulker 1984)

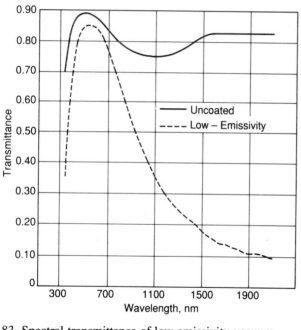

83. Spectral transmittance of low emissivity vacuum
coated glass in the visible and near infrared

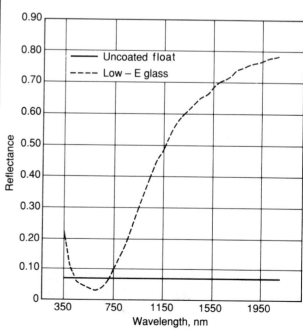

84. Spectral reflectance of low emissivity vacuum
coated glass in the visible and near infrared

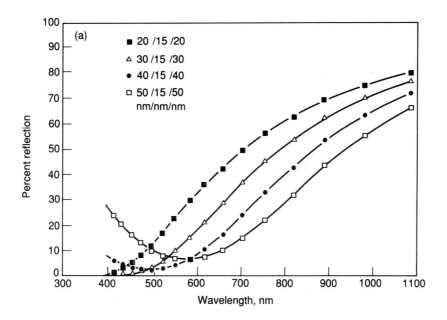

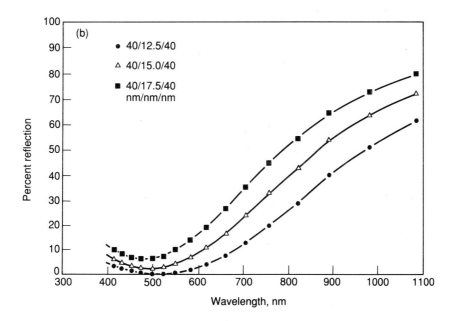

85. Effects of (a) oxide layer and (b) silver layer thicknesses on the spectral reflectance of low emissivity vacuum coated glass (20/15/20 signifies a coating in which a layer of silver 15 nm thick is sandwiched between two layers of tin oxide, each 20nm thick)

These coatings require careful handling after manufacture. They scratch more easily than the alternative, pyrolytic low-emissivity coatings (described later) and are more readily attacked by impurities in the atmosphere. However if reasonable care is taken and the coated glass is made into double glazed units within a matter of days after coating, very satisfactory yields are obtained.

It is interesting to note, however, that a far more durable $SnO_2/Ag/SnO_2$ coating can be made in a plant equipped with rotatable cathodes. These permit the sputtering of a hard protective layer of silicon or silica on top of the coating. The resulting product is claimed to ba as durable as the pyrolytic coatings.

Solar control coatings

A wide range of these coatings is on the market, mainly for architectural use. Although technically their function is to absorb or reflect solar radiation and to reduce glare, questions of aesthetics affect the choice of material since they greatly affect the appearance of a building in which they are used extensively.

Manufacturers rarely disclose the nature of the coating material used and one has to rely on limited information scattered through the literature. The most frequently used coating materials appear to be titanium oxynitride and an iron–nickel–chromium alloy used either alone or in combination. Such coatings would be expected to have good mechanical durability and resistance to atmospheric attack. Evaporated gold films have also been used, but they require mechanical protection, e.g. by placing them inside a laminated window.

Oxide and nitride coatings are made by reactive sputtering. For example, a tin oxide layer is produced using a pure tin source plate, a controlled flow of oxygen being fed with the argon into the coating chamber. The flow and other conditions must be regulated so as to ensure the formation of a stoichiometric coating on the glass but without forming an insulating layer of oxide on the target.

7.5. Chemical vapour deposition (CVD) and spray pyrolysis (Blocher 1981; Pulker 1984)

7.5.1. Process fundamentals

CVD is widely used for producing coatings on many materials. In the semi conductor industry it is used in making pure polycrystalline silicon for subsequent single crystal growth, for thin film epitaxial deposition and for the production of dielectric films (Bean 1981). A large part of the fundamental understanding of the process has been gained in work related to the processing of semiconductor materials.

A carrier gas transports reactive chemical vapours over a heated substrate on which chemical reactions take place and deposition of the coating occurs. The carrier gas may or may not take part in the reaction. Thus when depositing silicon, an inert carrier gas may be used, the coating being produced by the thermal deposition of silane. Alternatively a silicon coating can be produced by the hydrogen reduction of $SiCl_4$, the hydrogen acting both as a carrier gas and as a reactant.

$$SiCl_4 + 2H_2 \rightarrow Si + 4HCl$$

The rate of deposition depends on both kinetic and thermodynamic factors. These include (1) the equilibrium composition of the gas mixture, as determined by the free energy change of the reaction at the operating temperature (2) the composition of the gas mixture fed to the coating vessel (3) the kinetics of the processes which take place at the surface being coated and (4) the rates of diffusion of the reactants in the gas phase to the surface being coated and of the reaction products away from the surface.

At low temperatures, transport through the gas phase is rapid relative to the surface reaction rates. The concentration gradients perpendicular to the surface are low and the process is said to be kinetically controlled. At higher temperatures, the diffusion of reactants and reaction products through the gas phase become rate limiting and the concentrations of reactants and reaction products tend towards equilibrium. The process is then much less dependent on temperature and is said to be diffusion limited. Fig. 86 (p. 90) shows the existence of the two regimes for the deposition of Si by hydrogen reduction of $SiCl_4$.

Detailed accounts of the kinetics of CVD processes have been given by Eversteijn *et al.* (1970) and Wahl (1977).

7.5.2. Doped oxide coatings on glass

Two important applications of CVD for coating glass are referred to elsewhere in this book. One is the deposition of doped silica layers when making preforms for communication fibres (Chapter 10). The other is the application of tin oxide coatings on glass containers (hot end coatings) to provide a wear resistant surface (Chapter 5). This section will be concerned with the deposition of coatings having controlled electrical and optical properties. The applications include:

1. Coatings of high reflectivity (low emissivity) in the infrared for use in double glazing and to increase the energy efficiency of sodium discharge lamps.
2. Coatings of low electrical resistivity to make visual display devices.

The coating of flat glass with doped SnO_2, low-emissivity coatings has been a commercial process for several years. Both CVD and spray pyrolysis methods have been used. In both methods, the coating may be applied on-line between the float bath and the lehr. It is important to ensure uniform conditions over the full width of the ribbon, i.e. uniform temperature and (in CVD) uniformity both in the flow of coating gas to the surface and in its extraction.

Fig. 87 (Kalbskopf 1981) shows one type of nozzle which has been proposed for CVD coating. One gas stream consists of a mixture of the nitrogen carrier gas with $SnCl_4$ and the other of nitrogen of controlled humidity. The gases expand in the first chamber of the nozzle and are then accelerated to the 0.5 mm wide slit. The reactive vapours emerge as laminar streams and mix only by diffusion. The hydrolysis reaction takes place below the nozzle slit, partly on the hot glass surface and partly in the vapour phase close to it. The waste gases are evacuated at a rate low enough so as not to disturb the laminar gas flow below the nozzle. A problem of haze in the coating was reduced by the addition of a small percentage of HF to the gas mixture. Kalbskopf notes that the process is less sensitive to variations in the chemical parameters than to the control of the gas flow and the design of the deposition device.

The early material manufactured in this way for use in double-glazing was not as attractive in some respects as that made by magnetron sputtering. The effective emissivity was about 0.2, compared with 0.1 for the sputtered coating and the glass often showed a slight and variable haze. However the CVD coatings now available are considerably improved in both respects. A major advantage of the oxide coatings produced by either CVD or spray pyrolysis is that they are much more durable mechanically and chemically than the sputtered coatings and do not therefore require so much care in storage and handling.

Spray pyrolysis (Viguié and Spitz 1975) is also a widely used technique for applying oxide coatings to glass. A fine spray of a solution of the coating material is directed at the hot glass surface. The solvent evaporates to a greater or lesser extent before each droplet strikes the glass surface. The coating is built up from reaction products formed either in the gas phase or at the surface. Coatings may be applied to the glass either in an open or semi-enclosed chamber. Higher quality coatings are produced if the spray is broken down into smaller droplets by an ultrasonic atomiser (Manifacier *et al.* 1981, Blandenet *et al.* 1981)

Frank *et al.* (1981) give a detailed account of the use of tin-doped indium oxide films as IR-reflecting coatings in sodium vapour discharge lamps and incandescent lamps. Figs. 88 and 89 show the two types of lamp .

In the sodium lamp, a significant fraction of the energy consumed does no more than keep the discharge tube hot enough to maintain the required sodium vapour pressure. It makes no contribution to the light output. In earlier lamp designs, heat losses were reduced by surrounding the discharge tube with a Dewar flask. In the current design shown, the Dewar is replaced by a borosilicate glass tube, spray coated on the inside with a doped indium oxide film. In the incandescent lamp, the film is applied to the inside surface of the front lens and has the function of reducing the considerable amount of heat radiation which would otherwise be emitted from the lamp.

7.5.3. Conducting coatings

The doped oxide coatings are good electrical conductors with resistivities of only a few ohms per square. They have many uses based on this property, e.g. as transparent electrodes and in the electromagnetic screening of computer terminals for security reasons.

The electrical and optical properties of oxide films produced by both CVD and spray pyrolysis have been studied in considerable detail and the relevant physics is well understood (Köstlin *et al.* 1975; Jarzebski band Marton 1976a, 1976b; Chopra and Pandya 1983). The film materials are oxide semiconductors, the properties being controlled by doping with altervalent cations or anions (SnO_2/F, SnO_2/Sb or In_2O_3/Sn). Fig. 90 shows that the emissivity of the coating, measured near room temperature, is related to its electrical conductivity in a simple way. The results, for various levels of doping, lie close to the theoretical curve. For a low emissivity, the film resistivity should be of the order of 10 ohm/sq.

Transparent electrodes are essential components in liquid crystal displays, widely used in watches and pocket calculators. There are however some materials and processing problems which affect the stability of transparent electrode coatings, especially if they are applied to soda–lime glasses. Reactions involving alkali ions in the glass surface give rise to haze in the coating, but more seriously alkali contamination of the coating affects its electrical properties and may, in liquid crystal displays, poison the display material itself. To reduce these problems, techniques have been developed in which a barrier layer is applied before the coating itself is applied to prevent the migration of the sodium into the coating. The problem is greatly reduced if a borosilicate glass of low alkali content is used as the substrate.

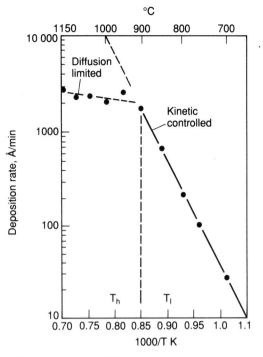

86. Kinetics of the CVD deposition of silicon by hydrogen reduction of SiCl₄
(Bean 1981)

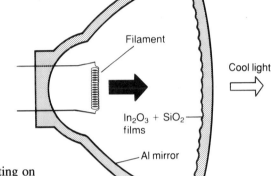

87. Cross-section through CVD nozzle for depositing doped SnO_2 coatings continuously on float glass
(Kalbskopf 1981)

89. Incandescent lamp with IR reflecting coating on front lens
(Frank et al. 1981)

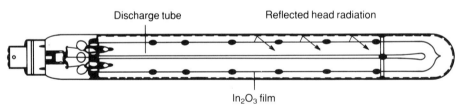

88. Low presssure sodium discharge lamp with IR reflecting jacket
(Frank et al. 1981)

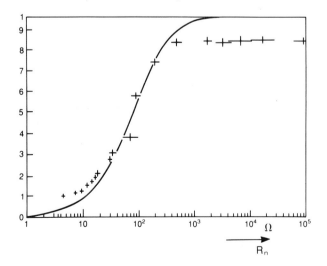

90. Dependence of thermal emissivity at 80°C on the surface resistance, R, of In_2O_3 . Various dopant concentrations
(Frank et al. 1981)

Most LCD's are used in situations where weight and space are at a premium. Drawing techniques have therefore been developed to produce glass substrate sheet with smooth, flat surfaces and thicknesses as low as 0.5 mm. They involves stretching thicker sheet at a relatively high viscosity (10^6 to 10^{10} Pa s) such that surface tension forces do not cause the width of the sheet to shrink (Wada 1987).

7.5.4. Electrochromic cells

The LCD effect is due to the orientation of the liquid crystals by the applied field. This increases the light scattering of the affected part of the layer. The response is fast, but visibility is not good in bright sunlight or at viewing angles greater than 60°.

The electrochromic effect may provide an alternative for some display applications. Also some electrochromic materials are cheap enough for it to be sensible to consider developing large window-sized cells ('smart windows') to control, at any required level, the light and solar energy entering a room. There are many papers covering the range from basic physics to descriptions of device performance, the following being only a sample list (Goldner and Rauh 1983; Baucke 1985; Svensson and Granqvist 1985; Estrada et al. 1988; Gambke and Metz 1989; Silver 1989).

In an electrochromic cell, the transmission of the active material in the visible is reversibly reduced by the application of a d.c. field, which causes electrons and charge-balancing cations to enter the material. The colour change persists even when the voltage is switched off, but the original transmission can be restored either by short circuiting the cell or by applying a reversed voltage.

A wide range of electrochromic materials exists, including many transition metal oxides and a number of organic materials. The most widely studied inorganic material is tungstic oxide and one of the most promising groups of organic materials is the lanthanide phthalocyanines.

Fig. 91 is a schematic representation of a tungstic oxide cell. Application of the voltage in the direction shown drives cations from the electrolyte E into the layer and an equal number of electrons enter it from the conducting coating. The cell reaction is therefore:

$$WO_3 + x(M^+ + e^-) = M_xWO_3$$

To incorporate the cations, some of the W ions change their valency from 6 to 5. Typically the transmission is reduced from 80 to 10%.

An important parameter, when considering applications, is the coloration efficiency, i.e. the reciprocal of the number of coulombs per unit area required to produce the change in absorption. This is wavelength dependant and for tungstic oxide is about 130 cm^2/C at 800 nm. The value for other oxides is much less.

Gambke and Metz give an interesting account of some of the problems which must be overcome if electrochromic cells are to be applied, especially in smart windows. The conductivity of present ITO coatings on glass obviously limits the flow of charge into the cell. They suggest that it is much too low to allow acceptable switching times for any large window. One also has to consider safety aspects. Presumably a smart window would need to meet safety glass specifications. Release of an acid electrolyte on fracture of a window would not be acceptable. This is not too serious a problem since cells have been described with solid electrolytes, which are either polymers or inorganic compounds. However no cell has yet been developed with the fragment retention properties of the well-known laminated safety glass.

Encouraging trials have however been carried out by Schott of an automatic dimming rear view mirror. A photocell picks up the headlights from a following car and the output signal is used to dim the reflection in the electroluminescent mirror.

Although the problems are large, they do not appear to have deterred many laboratories throughout the world from continuing the attempt to exploit this interesting application.

7.6. Sol-gel coatings (Mukkerjee and Phalippou 1985)

The process of sol–gel coating was developed many years ago by Geffcken (1951) and Schroeder (1969) of Schott Glaswerke. Most of the commercial development of the process has taken place within that company and has been described in a number of papers by Dislich and his colleagues (e.g. Dislich and Hussmann 1981; Dislich 1984; Arfsten 1984; Hinz and Dislich 1986; Dislich et al. 1989).

The main features of the coating process are illustrated in Fig. 92.

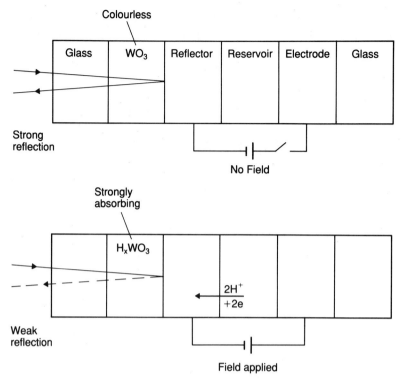

91. Principle of action of Schott electrochromic mirror. Application of a dc field drives hydrogen ions into the tunsic oxide layer, which becomes strongly absorbent thus reducing the intensity of the light reflected

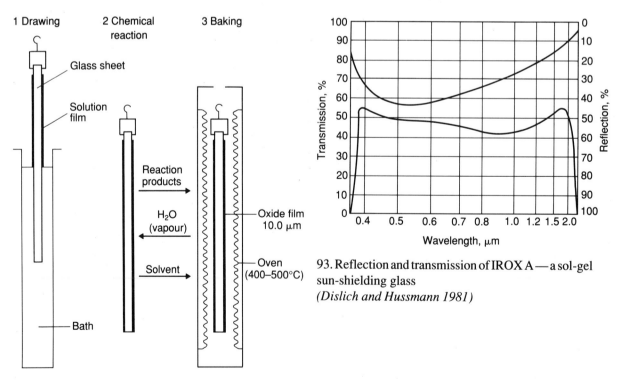

92. Coating glass sheets by the sol-gel dipping process
(Dislich et al. 1989)

93. Reflection and transmission of IROX A — a sol-gel sun-shielding glass
(Dislich and Hussmann 1981)

Large panes of float glass (up to 3×4 metres) are carefully cleaned and then dipped into a bath or a series of baths each containing a solution of a metal alkoxide in an alcohol–water mixture. The pane is withdrawn at a controlled speed. This speed, together with the viscosity of the solution, determines the thickness of the liquid film on the pane and hence the thickness of the coating eventually formed.

Hydrolysis and condensation of the sol film on the glass occurs by reactions of the kind described in Chapter 9 which results in its conversion to a solid gel. The panes are then heated to $400-500\,^{\circ}C$ in order to drive off volatile constituents (water and alcohol) and to harden the film. The thickness which can be applied in a single coat and bake is limited to about 0.5 micron.

Given careful attention to the relevant chemical factors, it is possible to prepare stable coating solutions which will form multi-component oxide coatings such as magnesium aluminium spinel and ITO (In_2O_3–SnO_2) (Arfsten 1984; Puyané and Kato 1983). Multiple layers can easily be produced by using a sequence of baths, one for each coating layer. In this way a wide range of optical properties can be obtained. By mixing solutions of the alkoxides of TiO_2 and SiO_2, films can be made with any required refractive index in the range 1.45 to 2.25.

The major commercial application of the process has been in the manufacture of a solar control glass product, IROX, the coating consisting of TiO_2 containing a colloidal dispersion of palladium. Fig. 93 shows transmission and reflection curves of one product in this range. Other standard products include interference filters for heat mirrors in tungsten halogen lamps, anti-reflection mirrors and colour filters.

A number of other applications have been suggested in the literature. They include low expansion coatings for strengthening glass, decorative coatings, and organically modified coatings (Schmidt 1988). Plates of controlled refracting properties have been made by using a microprocessor to vary the rate of withdrawal from the sol and so vary the film thickness (Hewak and Lit 1989).

It is clear from the applications developed so far that sol–gel coating is a very versatile technique and is likely to find many important applications in the future.

Chapter 8

Glass-Ceramics and Related Materials

8.1. Introduction

As indicated in Chapters 1 and 2, controlled microstructures on a fine scale can be produced in glasses by heat treatment. They may result from metastable immiscibility or from the nucleation and growth of a crystalline phase. This chapter deals with a number of materials of this kind. The types of microstructure and the range of properties that can be obtained is very large. They depend primarily on the composition of the glass, on the presence of nucleating agents and on the heat treatment given. When making some materials, additional control of the microstructure may be obtained by UV irradiation given before the heat treatment or by chemical treatment given afterwards. Commercially, the most important of these materials are the glass-ceramics.

8.2. Glass–ceramics (McMillan 1979; Beall and Duke 1983; Beall 1985, 1986; Strnad 1986)

8.2.1. General

The term 'glass–ceramic' covers a wide range of materials made by the heat treatment of a glass to convert it practically completely and in a controlled manner to a dense micro-crystalline body. Full advantage can be taken of the wide range of glass-fabrication processes available to produce the shapes required. The properties of the final material are determined by the crystallite size and by the composition and properties of the crystalline phases and the glass which have been produced by the heat treatment. The applications of these materials are widespread and are increasing. They range from domestic cooker tops and kitchen ware to telescope mirrors and laser gyroscopes.

The French chemist Réaumur is credited with the discovery of glass–ceramics in 1739. His materials were produced by heat treating soda–lime–silica glass. Later an attempt to commercialise the material was made by Gauchy who produced an artificial stone, used to pave some of the streets of Paris (1896). This early work had little practical impact and the current study and application of glass–ceramics all stems from the work of S. D. Stookey and his colleagues at Corning Glass Works beginning in the early 1950s.

The heat treatment schedules are critically important since they determine the texture (grain size and shape) of the material and, to some extent, the crystal phases which are produced. The type of schedule used is shown in Fig. 94 (p. 96). The early part of the schedule is very brief and does not usually greatly affect the final properties. Note however that in making certain types of seal to metal, the glass is not fully cooled and then reheated, as is implied by this diagram. It is converted to glass–ceramic as part of the sealing process.

The first isothermal stage at T_n produces the nuclei on which the crystal phases will grow. The nuclei are very small and very numerous and should be uniformly dispersed throughout the material. After nucleation, the temperature is raised to the second isothermal or growth stage at T_{gr} to produce crystals of the required phase and of the required size. The most commonly used nucleating agents are noble metals and oxides of cations of high field strength, which have limited solubility in silicate melts, e.g. TiO_2 and P_2O_5. The structural processes which occur in nucleation and crystal growth are complicated, differing considerably from one system to another. They have been, and still are, the subject of intensive research. Work in this field has been well summarised in the general references given above.

8.2.2. Types of glass-ceramic

It is convenient to follow Beall's classification (1985) of the various types of material now available.

a. Glass-ceramics based on β-quartz solid solutions.
  Compositions: a wide range in the SiO_2–Al_2O_3–Li_2O–MgO–ZnO system.
  Nucleating agents: ZrO_2, TiO_2, Ta_2O_5.
  Properties: Very low expansion coefficients, some zero or even negative. Some compositions are transparent.
  Applications: cooking ware, telescope mirror blanks, IR-transmitting cooker hobs.

b. Glass-ceramics based on β-spodumene solid solutions.
    Compositions: range from $Li_2O \cdot Al_2O_3 \cdot 4SiO_2$ to $Li_2O \cdot Al_2O_3 \cdot 10SiO_2$, with additions – especially MgO.
    Nucleating agent: TiO_2.
    Properties: similar to the β-quartz type – low expansion and very good thermal shock resistance.
    Applications: cooking ware, hot plates, ceramic regenerators for turbine engines.

c. Glass-ceramics based on cordierite.
    Compositions: SiO_2–Al_2O_3–MgO–TiO_2.
    Nucleating agent: TiO_2 – about 10%.
    Properties: Good mechanical properties, depending on heat treatment. Fracture toughness can be as high as 2.5 MPa $m^{1/2}$, about three times higher than most silicate glasses.
    Moderately low expansion coefficient.
    Excellent microwave dielectric properties.
    Applications: Widely used in the US for radomes.

d. Glass-ceramics based on lithium disilicate.
    Compositions: Mainly Li_2O and SiO_2 (see Strand p.77).
    Nucleating agents: noble metals, P_2O_5.
    Properties: Good mechanical properties and surprisingly good electrical properties in spite of their high alkali contents.
    Applications: Photosensitive materials, chemically machinable materials for plasma display panels, fluidic devices etc.

e. Machinable fluormica glass–ceramics.
    Compositions: based on fluorine phlogopite ($KMg_3AlSi3_3O_{10}F_2$), subpotassic phlogopite ($K_{1-x}Mg_3A_{1-x}Si_{3+x}+O_{10}F$) and other naturally occurring micas.
    Nucleating agents: Self-nucleating.
    Properties: Fine grain structure consisting of randomly oriented platelets. Precision machining possible using metal working tools.
    Applications: precision electrical insulators, microwave tube windows, dental restorations and other prosthetic applications.

f. Glass–ceramics based on pyroxenes and pyroxenoids (basalts and slags).
    Compositions: natural basalts and slags.
    Nucleating agents: Fe_2O_3, Cr_3O_3, sulphides and fluorides.
    Properties: good chemical durability and wear resistance. Intensively studied both in Eastern Europe and the UK (see Strand loc cit).
    Applications: Building products e.g. floor tiles. Conveyor chute linings.

g. Glass–ceramics based on canasite and agrellite.
    Compositions: Canasite is $Na_4K_2Ca_5Si_{12}O_{30}F_4$ and Agrellite is $NaCa_2Si_4O_{10}F$.
    Nucleating Agents: excess fluoride.
    Properties: blade-like, interlocked crystals, high strength and high fracture toughness.

h. Glass–ceramic paper based on fluorhectorite.
    Compositions: Fluorhectorite is $LiMg_2LiSi_4O_{10}F_2$.
    Properties: The material swells when immersed in water and the fine mica platelets form a stable suspension.
    Application: ceramic paper and board. Will withstand 500°C.

i. Devitrifiable solder glasses.
    Compositions: PbO–ZnO–B_2O_3 system.
    Properties: Expansion match to TV tube and similar glasses (α ca. $90 \times 10^{-7}/°C$)
    Applications: sealing together components of TV tubes and semiconductor packages.

j. Compositions for metal coating.
    Compositions: various, depending on application.
    Applications: acid resistant for lining chemical reaction vessels, coatings for metal printed circuits boards.

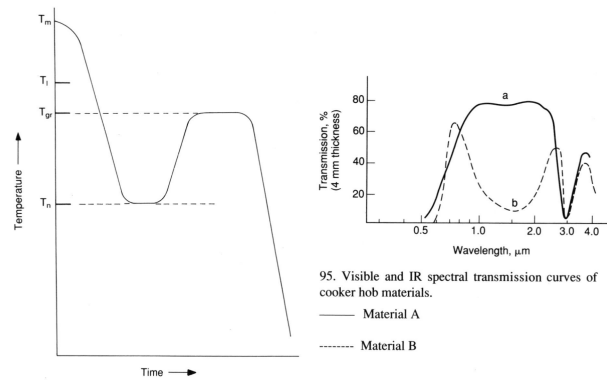

94. General form of the temperature time schedule used in making a glass-ceramic

95. Visible and IR spectral transmission curves of cooker hob materials.

—————— Material A

-------- Material B

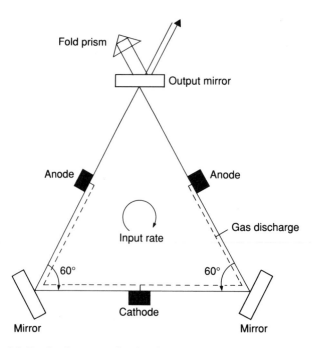

96. Basic elements of a ring laser gyroscope
(Hiller 1983)

The list gives a broad outline of the range of application of glass–ceramics. The following section looks at some of the these in more detail.

8.2.3. Applications of glass–ceramics

Low-expansion materials
Since 1967, low expansion, lithia–alumina–silica glass–ceramics have been widely used to make electric cooker hob plates. A considerable amount of research has been carried out by the manufacturers on the material itself and on its performance in appliances (Scheidler 1974; Scheidler and Kristen 1980; Scheidler and Taplan 1984). The plates must be able to withstand the quite severe maltreatment of normal domestic use. The relatively high strength of about 100 MPa combined with the low expansion coefficient makes it possible for a plate running at 650 °C to withstand the thermal shock of cold water being poured onto it. Stability of properties is an important factor and it has been shown that no changes are detectable after heating the material for 6000 h at 700 °C.

Although the heating element is mounted beneath the plate, a significant amount of heat is transmitted to the pan or casserole by radiation. With the introduction of tungsten–halogen lamps as heating elements in some models of cooker, it has become necessary to modify the material composition to give high transmission in the near infrared but to reduce transmission in the visible to eliminate glare. Fig. 95 compares the transmission curves of two commercially available hob materials, that labelled 'a' being designed for the new heating elements.

A similar material is used to make Corning's glass–ceramic cookware 'Visions'. This too is transparent, so when used in conjunction with a transparent hob material, the food is heated directly by radiation rather than by conduction and convection as in traditional cooking. (The final result will still depend mainly on the cook!).

A very low expansion coefficient is obviously important in types of optical equipment for which high dimensional stability is important. Two examples are worth mentioning. A number of large optical telescopes have been built in recent years with low-expansion glass–ceramic mirrors. The castings measure up to 8 m in diameter – amongst the largest pieces of glassware made (Schuster 1989). Lindig and Pannhorst (1985) have investigated some of the factors, especially thermal cycling, which affect the dimensional stability of this material.

The second application is in the laser gyroscope (Fig. 96, Hiller 1983). The laser body is made from a cast block which, after heat treatment, is shaped by cold-working operations (cutting, grinding, polishing). The He–Ne discharge generates radiation which propagates in opposite directions round a triangular path. If the gyroscope rotates as shown, there is a difference in frequency between the beams, and the rate of rotation can be measured by photo-electrically counting the interference fringes. For a cavity with a perimeter of 450 mm, one count corresponds to an angular rotation of about 1.5 seconds of arc. Even the best low-expansion material is not quite good enough to maintain the stability required so it is necessary to compensate for dimensional changes due to thermal expansion by a feedback control which adjusts the position of one of the corner mirrors.

The vacuum-tight seals which attach the electrodes and mirrors must be made by low temperature techniques so that the block is not distorted. Seals are made either by ringing together very flat and highly polished surfaces or by the use of indium gaskets. In the latter method, the surfaces to be joined are first gold coated. Indium wire or foil is placed between the seal surfaces and the assembly is heated in vacuo to melt the indium.

Glass ceramics based on basalt and metallurgical slags
Useful summaries, in English, of work in Eastern Europe on these materials have been given by Hlavac (1983), Strnad (1985) and Sarkisov (1989). Major constituents are CaO, MgO, Al_2O_3 and SiO_2. Depending on the source of material, additives are made to the batch to maintain a reproducible composition. Materials used as nucleating agents include Cr_2O_3, sulphides and fluorides. The melt may be continuously rolled into sheet or centrifugally cast into pipes. Although having only moderate strength, the abrasion resistance is very good and the materials have been widely used as floor tiles and for lining rock conveyors. Sarkisov describes recent developments to produce materials of more attractive appearance (synthetic granite, artificial marble) for use in building facades. Work in the UK on similar materials has been described in detail by Davies *et al.* (1970).

At present there does not appear to be a strong interest in western countries in producing and marketing these materials. This may grow if economic incentives are offered for disposing of unsightly slag heaps. (It may be worth mentioning here that there is considerable current interest in disposing of hazardous asbestos waste by converting it into a glass. Perhaps this material could be converted into a glass ceramic.)

Inorganic paper and board (Garfinkel 1986)
Paper and board with good electrical and thermal properties at elevated temperature can be made by reprocessing a mica-containing glass–ceramic. The aim of the development has been to produce a grade of paper which, in addition to having the desired physical properties, also has a smooth surface finish, superior in this respect to other

inorganic papers. Various compositions have been used, but the favoured materials contain a major proportion of lithium fluorhectorite, $LiMg_2LiSi_4O_{10}F_2$. When the glass–ceramic is exposed to water, the mica phase disintegrates into very small platelets of high aspect ratio. The aqueous sol may then be flocculated by introducing cations which are larger than the sodium and lithium cations normally present on the platelet surfaces. Alternatively organic flocculating agents may be used. The flocculated material is processed using commercial paper-making equipment. Other processes have been developed to make board or foam from the flocculated sol.

Seals to glass–ceramics

Making vacuum-tight seals between metals and glass–ceramics presents more problems than when sealing to glass. There is a considerable change in expansion coefficient when the parent glass is heat treated. The process must be carefully controlled to ensure that the expansion coefficient of the final glass–ceramic is a good match to that of the metal. In spite of the obvious difficulties, McMillan and his colleagues have shown that satisfactory seals can be made on a routine basis (McMillan *et al.* 1966a, 1966b).

Two sealing methods are used. In one, the glass–ceramic is formed in the normal way and is then sealed to a matching metal component using one of the well known ceramic–metal sealing techniques, i.e. by using an 'active metal' vacuum brazing technique or by using a thin glaze layer.

In the second method, the parent glass is cast into a mould in which the metal component has been positioned. The assembly is then subjected to the crystallising heat treatment. It is, of course, essential that the heat treatment should produce a glass–ceramic which has a good expansion match to the metal. The seal is allowed to cool only when the crystallising heat treatment is complete.

Machinable glass–ceramics

Beall (1971), Grossman (1972) and Chyung *et al.* (1974) have described a group of machinable glass–ceramics in which the main crystalline phase is a fluorphlogopite mica. The microstructure consists of a mass of small, interlocking mica crystals. These materials are machinable to close tolerances by normal metal-working methods. The machinability is due to the easy cleavage of the mica crystals combined with their random orientation which acts to blunt and deflect the small cracks formed in machining.

The potential range of application of machinable glass-ceramics has been extended by Vogel *et al.* (1986) who have developed a material for prosthetic applications. They added CaO and P_2O_5 to the glass so that the crystallised material contains an apatite phase in addition to the phlogopite. The new material was found to be bioactive, i.e. implants were shown to bond to adjacent bone in tests on animals.

Glass–ceramic coatings for thick-film hybrid circuit boards

Thick-film hybrid circuits are commonly made by the screen printing of inks, which form the circuit components, onto flat alumina ceramic plates. After printing, the plates are fired at a temperature of about 850°C which volatilises the solvent and decomposes the inks to form the required conductors, resistors, etc. Units of this type become increasingly difficult to make, more fragile and more expensive as their size increases.

For some years there has existed a need to make relatively cheap, rugged boards with sizes of the order of the standard 'Eurocard'. It has been recognised that an attractive possibility would be to print the circuit onto a vitreous enamelled metal plate. A major difficulty was that the enamels available had too low a softening temperature to be compatible with the standard inks at the firing temperatures used to apply them. Ink manufacturers were unwilling to develop special low firing inks to meet the uncertain future requirements of the new type of board. Another problem was that the enamels contained alkalis which migrate under the action of the local electric fields, with a deleterious effect on circuit performance. Also the enamels had originally been developed for application to a conventional enamelling steel, which has a relatively poor oxidation resistance. It was therefore necessary to coat both surfaces of the steel, thereby limiting the possibility of applying effective cooling to the back surface (Jones 1985).

A very thorough investigation of all the problems involved was carried out by a team at the RCA laboratories (1981). They introduced the idea of using a glass–ceramic coating applied electrophoretically, a technique capable of producing a very uniform coating. The coating material was a composition in the $MgO–BaO–B_2O_3–SiO_2$ system, thus avoiding the problem of alkali migration. It was claimed that the coating was sufficiently refractory to withstand the firing on of inks at temperatures as high as 900°C, suggesting that the range of standard inks might be used. The steel used by RCA still had a relatively poor oxidation resistance, and it was necessary to coat both surfaces of the board and to be especially careful to ensure that corners and hole edges were protected.

In the development described by Jones (loc cit), an alloy of high oxidation resistance was used (basically a Fe–Cr–Al alloy). During initial firing, this forms a strongly adherent aluminium oxide surface film which greatly reduces further oxidation. With this material, coating on one side only is sufficient. The metal has an expansion coefficient somewhat higher than that of enamelling steel (140×10^{-7}°C^{-1} from 20 to 700°C). The coating is a glass-ceramic in the system Li_2O– ZnO–SiO_2. In spite of the presence of Li_2O, the material is a good electrical insulator at the operating temperature (10^{11} ohm cm at 200°C – about 1000 times greater than the resistivity of the parent glass). The presence of the alumina oxidation film on the metal ensures excellent adhesion of the coating. As in the RCA work, some development was required to produce suitable inks but it appears that this system now meets the user's requirements.

In a parallel development, similar techniques are being applied to make heaters for various types of domestic appliance. Thus a spiral heating element can be formed by screen printing onto the surface of the glass–ceramic coating.

8.3. Microporous glasses

Chapter 1 refers briefly to the phenomenon of *sub-liquidus* immiscibility and to the work, beginning in the mid-1960s, which led to an understanding of the thermodynamic and kinetic factors that control the microstructure and to the current recognition that *sub-liquidus* immiscibility occurs in many oxide glass systems.

Nearly thirty years earlier Corning Glass Works had exploited the effect commercially in the manufacture of 'Vycor' glass (Nordberg and Hood 1938). A sodium borosilicate glass, having a composition within the shaded area in Fig. 97 (preferably close to the line OL) is heat treated to cause phase separation.

The microstructure produced is similar to that shown in Fig. 8b, in that it consists of two interpenetrating phases, one of which is almost pure silica whilst the other is an acid-soluble sodium–borate glass. After dissolving out the borate phase, a silica skeleton is left having a very small pore size (20–40 AU). In the 'Vycor' process, the porous material is dried and finally sintered at 900–1200°C to produce a material very similar in properties to vitreous silica. Volf (1961) has given a detailed account of the various stages of the process.

'Vycor' glass is widely used, e.g. as a lamp envelope material, but this section will be concerned only with applications of the microporous silica skeleton material, produced by acid leaching the phase-separated glass.

The pore structure and the internal surface area may be varied by varying the heat treatment . Typically the latter may be as high as 200 m^2/gm. The ability of the internal surface to act as an adsorbant for a variety of substances may be controlled by appropriate chemical pretreatments.

Messing (1977) describes how various enzymes are immobilized by adsorption onto a silane-treated skeleton and how materials so produced may be used in biochemical analysis and in food processing. Janowski and Heyer (1982) give a very full account of the preparation, properties and applications of microporous glasses. Amongst the applications they describe are catalyst supports, desiccants, molecular sieves and chromatographic materials.

The application of tubes of microporous glass for desalinating sea water by reverse osmosis is described by Phillips *et al.* (1974). Fig. 98 illustrates the principle of the process. By applying to the left hand chamber a pressure P greater than the osmotic pressure π, pure water is caused to flow through the membrane into the right hand chamber, against the direction of the osmotic pressure. Organic membranes are already used successfully to purify brackish water but are subject to bacterial attack when working in sea water. It was shown that the glass membranes had an output of 0.25 m^3/m^2 per day, less than that of the best organic membranes, but the superior durability of the glass membranes gives them the advantage in some applications.

Recent developments in the applications of microporous glasses are described by Schnabel and Langer (1989).

8.4. Colloid colours (Weyl 1951; Bamford 1977)

A number of glasses coloured by colloidal dispersions of Cu, Ag, Au, CdS and CdSe are used as signal and filter glasses (Chapter 2). Close control of the colour is essential and this is achieved by control of the glass composition, the melting schedule and the heat treatment of the glass. The latter controls the nucleation and growth of the colloidal particles.

Badger *et al.* (1939) have measured the absorption spectra of a gold ruby glass based on a soda–lime–silica composition after heat treatment for various times up to 9 h at temperatures ranging from 575° to 900°C. The spectra were shown to be very similar to those of aqueous gold hydrosols. Fig. 99 shows the spectra of the glass after heating for various times at 600°C.

Doremus (1964) produced glasses containing gold particles of controlled size and showed that the measured absorption spectra were in good agreement with those calculated using available data for the optical constants of gold and Mie's equations for the absorption and scattering of light by colloidal particles. The absorption spectra of glasses coloured yellow or amber by colloidal silver were successfully interpreted in the same way (Fig. 100;

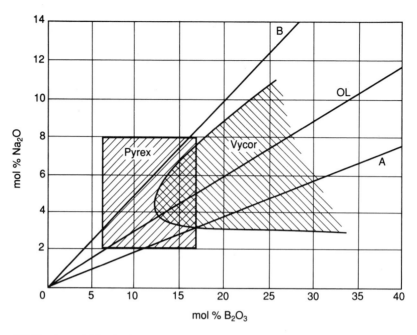

97. Region within the Na_2O-B_2O_3-SiO_2 system for making Vycor glass
(Volf 1961)

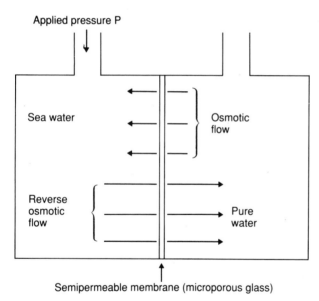

98. Purification of water by reverse osmosis.
Application of pressure P causes pure water
to flow from left to right
(Partridge and Phillips 1978)

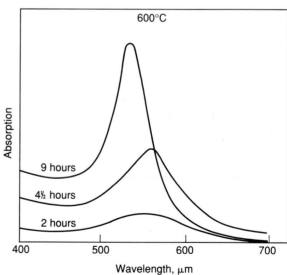

99. Striking of gold ruby at 600°C
(Weyl 1951)

Doremus 1965). Similar comparisons of measured and calculated absorption spectra have been made by Bamford (1976) for 'Spectrafloat' and by Rawson (1965) for a borosilicate glass stained using copper and silver salts.

Only small concentrations of metal are needed to produce the intensity of colour required, typically 0.01–0.1% for gold, 0.05–0.5% for silver and 0.1–0.5% for copper. Control of the melt chemistry is important for ensuring proper nucleation of the metal colloid. At the glass melting temperature, the metal is dissolved in the glass, i.e. it is present in the ionic form. Nucleation occurs at much lower temperatures and additions to the glass batch are needed to ensure that the metal ions are reduced to metallic form at the nucleation temperature. These are oxides of polyvalent elements, of which tin oxide appears to be the most effective.

8.5. Photochromic glasses (Megla 1966; Smith 1967; Araujo 1977, 1980, 1986; Gliemeroth *et al.* 1981)

A large number of photochromic materials are known but only a small number of photochromic glasses. The commercial materials, widely used in sun-glasses, all depend on the presence of a dispersion in the glass of submicron sized particles of silver chloride (Smith 1967).

Their ability to darken when exposed to strong sunlight and to recover their transmission in the shade is well known. They were discovered by Armistead and Stookey of Corning Glass Works (1964). Since then, much research has been devoted to the study of the glass chemistry and the physics of the darkening and bleaching processes.

Darkening involves a process similar to that which occurs when a photographic film is exposed. A sufficiently energetic photon interacts with a silver chloride particle decomposing it to metallic silver and a chlorine atoms, as shown in Fig. 101. The chlorine atom is held close by in the glass matrix and is available for recombination with the silver at a later stage (i.e. during bleaching). In borosilicate photochromic glasses, the AgCl is concentrated in the borate phase and this prevents the Cl atoms from diffusing away. Recombination requires either thermal energy or interaction with light of lower energy than that which caused the initial darkening.

Commercially there is considerable interest in developing glasses with more rapid response (especially more rapid bleaching) and current materials are significantly better in that respect than those which were available initially. There are no indications either from laboratory studies or practical use of any optical fatigue.

The glasses contain about 0.5% of silver and the particles are extremely small – about 100 AU in diameter, the spacing between the particles being about 600 AU. Thus the particles are much smaller than those in a slow, high resolution black and white film, which are about 1000 AU in diameter.

Smith (loc cit) presents interesting results of experiments to assess the possibility of using photochromic glasses for solar control glazing of buildings and automobiles. This application has not been pursued, presumably because of the more recent development of cheaper alternatives.

Consideration has also been given to using photochromic glasses as optical storage media. This idea also does not appear to have been followed up, presumably because the image is not sufficiently permanent for some applications and cannot be erased quickly enough for others.

8.6. Photosensitive glasses

If a small percentage of CeO_2 (ca 0.05%) is added as a sensitiser to the glass, nucleation of the metal colloid can be enhanced by irradiating the glass with ultraviolet light before heat treatment. By placing a suitable mask or negative between the light source and the glass, a permanent photographic image of excellent quality and high resolution is formed.

The cerium oxide is present in the glass in both the trivalent and tetravalent states. According to Stookey (1949) the mechanism of sensitisation involves the reactions:

$$Ce^{3+} + h\nu \rightarrow Ce^{4+} + e^-$$

$$Ag^+ + e^- \rightarrow Ag^\bullet$$

Heat treatment at about 450°C after exposure causes nucleation and growth of the silver, resulting in particles about 100 nm in diameter. Analogous processes occur in the copper and silver containing glasses. A useful account of the manufacture and processing of these materials has been given by Nebrensky (1965).

Photosensitive opal glasses have also been developed in which the colloidal metal particles are used as nuclei for the growth of crystals of sodium fluoride. The most recent development in the series of photosensitive glasses is a material which can develop the full range of spectral colours. This is a photosensitive opal in which the

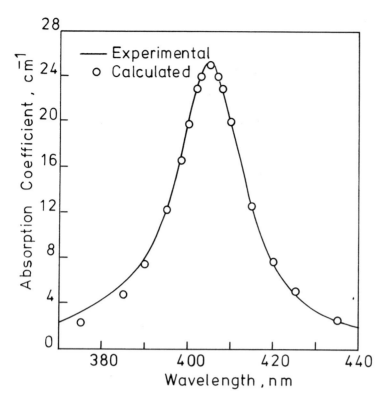

100. Absorption spectrum of silver-yellow glass due to photonucleated metallic silver particles *(Doremus 1966)*

$$n\,AgCl \xrightarrow{\text{h}\nu} n\,Ag^{\bullet} + n\,Cl^{\bullet}$$
$$\searrow$$
$$(Ag^{\bullet})_n$$

101. Reaction scheme *(Smith 1967)*

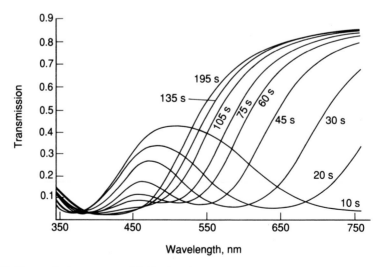

102. Transmission curves of fully developed polychromatic glass 1.5 mm thick for various first exposure times *(Beall 1978)*

irradiation and heat treatment affects the shape as well as the size of the fluoride crystals. To produce a multi-coloured image, the glass is given two UV irradiations, each followed by a heat treatment. The intensity of the colour depends primarily on the final exposure whilst the hue depends on the first (Stookey *et al.* 1978; Borrelli *et al.* 1979).

Table 15 shows a typical compositon and Table 16 an exposure/heat treatment schedule.

Fig. 102 shows transmission curves for the fully treated glass as a function of the first exposure time.

Table 15

Typical polychromatic glass composition
(Smith 1979)

	wt-%
SiO_2	60.0
Na_2O	15.8
ZnO	4.8
Al_2O_3	6.8
F^-	2.3
Br^-	1.0
Ag^+	0.01
CeO_2	0.05
Sb_2O_3	0.20
SnO	0.05

Table 16

The photochemical process to produce a polychromatic glass

1. First UV exposure (300 nm) (5 sec to 5 min)	Develops latent silver image in glass
2. First heat treatment 450–500°C	a. Sub – colloidal silver nuclei population density a function of 1.
500–550°C	b. Nucleated pyramidal NaF (Na,Ag)X crystallites
3. Second UV (300 nm) exposure (10 min to 2 h)	Latent silver image in crystallite apices
4. Second heat treatment 300–460°C	Colour image forms. Anistropic silver particles.

Microstructural studies have shown that many of the fluoride particles grown after the first exposure are elongated pyramids in contrast to the cubic crystals found in the normal photosensitive opal glasses. The anisotropy is a major factor in determining the colour (Smith 1979; Beall 1978).

8.7. 'Chemical machining' of glass ceramics

The theme running through this chapter is the development of materials having useful properties obtained by controlling the microstructure. Not all the materials have been widely used but there has been a progressive introduction of new ideas which have produced new effects. In such a programme a relatively small innovation may open up a large and important field of applications. This is illustrated in the present section.

When photosensitive glasses based on the system Li_2O–SiO_2 are irradiated and then heat treated, heterogeneous nucleation of either $Li_2O \cdot SiO_2$ or $Li_2O \cdot 2SiO_2$ occurs on the metal colloid particles. It was found that the crystalline phases are considerably more soluble in acid than the parent glass (Fig. 103). Thus by UV exposure through a mask, then heat treatment to crystallise the exposed areas and finally dissolution of the crystallised regions, one can produce a plate containing holes or grooves in positions exactly defined by the mask. Suitable compositions are given by Strnad (1986 p.77)

The materials have been used to make many devices, such as that shown in Fig. 72, a flat-panel monitor developed by Siemens (Schauer 1982) in which the spacing hole plate is made as described. The hole density is 781 per cm^2 and the monitor can display 28×80 charachters with a 9×16 dot field matrix for each character.

Fig. 104 shows a micro-lens assembly made using a similar material and with similar heat treatments but without the final step of acid solution (Borrelli *et al.* 1985). The material outside the circular regions is exposed and then crystallised. Volume contraction during crystallisation exerts pressure on the unexposed glass causing it to be extruded slightly from the plane surface of the plate. Surface tension produces the necessary curvature and a closely spaced array of lenses is thus formed in the plate. Such plates are of interest for use in small photocopiers.

There are many other applications for this material when glass and glass–ceramic components are required involving very finely detailed machining. It might be possible in some instances to produce the same components by traditional mechanical techniques but it would be expensive and the risk of breakage would be high.

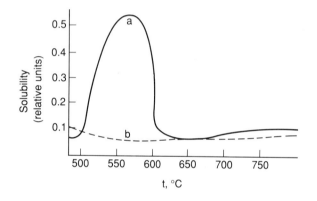

103. Change in the solubility in HF of a photosensitive glass based on Li$_2$O–SiO$_2$ after heat treatment at the given temperatures
(Strnad 1985)
(a) previously nucleated by irradiation — and now crystalline; (b) unirradiated — and still glass

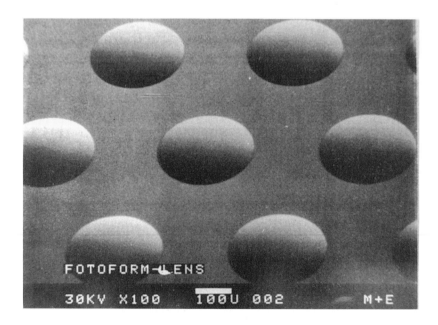

104. Part of a lens array made by controlled UV exposure followed by heat treatment
(Borrelli et al. 1985)

Chapter 9

Glasses made by Sol-Gel Processes

9.1. Introduction

The term 'sol–gel process' covers a number of methods for making a solid material, starting from:

> (a) a single-phase liquid (containing precursor compounds of the oxide glass components) which on reaction solidifies to a rubber-like gel or (b) a stable colloidal suspension of solid particles in a liquid medium. Chemical additions to the suspension initiate a sequence of chemical and physical changes which ultimately cause it to gel or solidify.

After further reactions, involving the elimination of water and/or alcohol, the material is densified by heating.

A sol–gel process has been used for many years to apply coatings of controlled optical properties to flat glass (Chapter 7). Much of the present very large research interest in sol–gel materials and processes was initiated by Dislich's account (1971) of how small *bulk* specimens of a borosilicate glass could be made in this way. The literature covering both the scientific and technological aspects of the process is extensive, much of it to be found in conference volumes edited by Gottardi 1982; Scholze 1984; Brinker *et al.* 1984a; Zarzycki 1986; Klein 1988a; and Sakka 1988b).

For making bulk glasses, the technique has a number of potential advantages:

1. Very pure raw materials are available.
2. Fully dense material can be made at temperatures much lower than those required in conventional glass melting. This helps to maintain purity by reducing crucible corrosion and the transfer of volatile material from the melting furnace.
3. When using method (a), the materials should, in principle, be very homogeneous since they are obtained by molecular mixing.
4. The chemistry can be tailored in such a way as to incorporate organic radicals in the material structure.
5. Materials of controlled porosity and high surface area can be made. Some have potential applications in that form. Others can be doped with chemically active or bioactive materials.
6. Fibres and films can be made directly from the sol and then cured.

When making bulk specimens, a major problem initially was that the gels cracked on drying. This problem has now been largely overcome and blocks of considerable size can be made. Even so the process is not a quick one. Depending on the particular method used and the sol composition, it may take several hours to reach the gel point and several hours more to remove the liquid phase.

9.2. Processes

9.2.1. Starting from alkoxides

A simple example of this class of process involves the hydrolysis of tetramethoxy– or tetraethoxy–silane (TMOS or TEOS) in a mixture of ethanol or methanol with water. The overall reaction involves hydrolysis followed by condensation:

$$n Si(OR_4) + 4n H_2O > n Si(OH)_4 + 4n ROH$$

$$n Si(OH)_4 > n SiO_2 + 2n H_2O$$

(This is an oversimplification of what actually occurs. There is in fact a sequence of intermediate reactions.) By working at a sufficiently high concentration of siloxane, the reactions result in the formation of a solid gel containing a high volume fraction of alcohol and water. The reaction rates are usually accelerated by adding either an acidic or a basic catalyst (e.g. HNO_3 or NH_4OH). The character of the resulting gel depends considerably on the type of catalyst used (Keefer 1984; Brinker *et al.* 1984b; Klein and Garvey 1984). Gels prepared with an acid catalyst normally require a very long drying time and have a large drying shrinkage. Gels prepared using a base catalyst shrink far less and have a much lower bulk density.

Stress builds up in the drying gel, caused by capillary forces which are directly proportional to the evaporation rate and inversely proportional to the pore size (Zarzycki *et al.* 1982). This may be so high as to fracture the gel. Two methods have been developed to reduce the problem. Hench adds to the sol a mixture of methanol and a drying control chemical additive (DCCA) – usually formamide. This has a much lower vapour pressure than methanol. The mixture reduces the rate of evaporation during the early stage of drying and so reduces the stress build up (Wallace and Hench 1984; Wang and Hench 1984; Orcel and Hench 1984). By this method, it was possible to produce, by acid-catalysed reaction, silica gel monoliths 100 cm³ in volume within a processing time of 2 days.

Problems still arise if any organic material remains in the gel when the final heat treatment begins. This may be converted to carbon or will produce fine pores if it oxidises.

Other methods of drying have been developed which remove the volatiles more completely and also greatly reduce the risk of cracking. The hypercritical solvent evacuation technique was introduced by Zarzycki and his colleagues (Zarzycki *et al.* 1982; Prassas *et al.* 1984).

The reaction mixture, contained in a mould, is placed in an autoclave with excess alcohol. This is pressurised with nitrogen. The temperature and pressure are increased above the critical point (ca. 240°C and 80 bars). When the hypercritical state is reached, the solvent is slowly vented over a period of several hours. The autoclave is then purged several times with nitrogen and is finally cooled.

9.2.2. Multi-component glasses

These can be made by adding to the sol the appropriate metal alkoxides or soluble inorganic salts, e.g. citrates, acetates, tartrates which decompose at a relatively low temperature (Thomas 1988). Amongst the alkoxides which have been used are sodium methoxide, aluminium secondary butoxide, trimethyl borate and double alkoxides such as $Ca(Al(OEt)_4)_2$. The nature and speed of the hydrolysis and condensation reactions in such mixtures are greatly affected by the composition of the sol. Some degree of de-mixing or chemical segregation may occur which will obviously affect the homogeneity and properties of the final glass.

Schmidt (1984) has described an interesting family of materials, the organically modified silicates, in which the sol–gel technique is used to incorporate organic groups into the structure of the material. Such materials obviously cannot be processed at the same high temperatures as the organic glasses. They should be regarded as a class of materials with their own potential field of applications.

The modification is achieved by incorporating, at the sol stage, compounds containing non-hydrolysable –Si–C– bonded ligands. The synthesis principle of one such material is illustrated in Fig. 105 (Philipp and Schmidt 1984).

105. Organically modified silicates. Introduction of organic groups into a silicate network during a sol-gel reaction
(Sckmidt 1988)

9.3. Glasses from ultrafine powders (Rabinovich 1988)

A process which has similarities to the sol–gel process described above has as its starting point very finely divided oxide powders, some of which are commercially available. These are incorporated into an aqueous or organic liquid medium. Additives cause the material to gel, after which it is dried and fired. At present, it is difficult to produce a wide range of compositions by this method and most of the work has concentrated on pure or slightly doped silica powder.

In the Rabinovich method, commercial 'fumed silica' powder made by the hydrolysis of $SiCl_4$ in an oxygen–hydrogen flame is dispersed in water. The sol is cast into moulds when it gels in 1 to 2 h. This gel proved difficult to dry without cracking. The water content was high (ca 60%) and the pores were small (ca 14 nm). The cracking problem was overcome by 'double processing'. The dried gel was ball-milled and then redispersed in water. This produces a significant fraction of larger pores (ca 500 nm) which provide channels for the escape of water. Fully dense silica glass could be made by sintering at 1350°C. Heating in a sequence of controlled atmospheres is necessary to obtain the best results – using a sequence such as the following:

1. Helium at low temperatures to ensure the elimination of porosity.
2. Chlorine to remove –OH groups and thereby improve the IR transmission.
3. Fluorine-containing gases or vapours to eliminate bubble formation on reheating.

In a similar process, developed by Scherer, the fumed oxide powder is suspended in chloroform, the sol being stabilised by an addition of 1–decanol which is adsorbed on the surfaces of the particles, so preventing agglomeration. The sol is made to gel by exposing it to ammonia vapour which reaches it through the porous walls of the container. The technique was also used, with appropriate dispersing agents, to make high density ceramic bodies (Al_2O_3, TiO_2) (Scherer 1984).

Clasen (1987 a,b) used the controlled hydrolysis of TEOS to make monosized silica particles. Compacts made from this material were successfully sintered to transparent silica glass. He also describes work on material made from fumed silica powder, using an approach similar to those of Scherer and Rabinovich. Finally Dorn *et al.* (1987) describe the preparation of low-loss silica fibre from material made by the hydrostatic pressing and sintering of mixtures of fumed silica and germania.

The considerable research activity on alternative methods for making high purity, high silica glass by methods closely related to the sol–gel process is an indication of the interest in manufacturing methods which may be faster and cheaper than the present chemical vapour deposition techniques.

9.4. Applications

Although many of the papers in this field describe work carried out with particular applications in mind, relatively few have reached the stage of production. The following account, though far from complete, should be sufficient to show the versatility of the process and the scope for a large range of applications.

Drying problems which cause difficulties when producing bulk material hardly exist when making fibres, films, coatings and thin-walled shells. Many of the more promising applications are in this area.

Sowman (1988) gives an interesting and detailed account of the production of refractory oxide fibres made from materials in the Al_2O_3–B_2O_3–SiO_2 system. In one method of fibre formation, a solution of the compound $Al(OH)_2(OOCCH_3) \cdot 1/3H_3BO_3$ in an aqueous silica sol is concentrated until the viscosity is of the order of 10 Pa s when it can be extruded through a spinerette. After firing at 1000°C, the fibres have a tensile strength of 1.04 GPa and a Young's modulus of 155 GPa. A range of fibre compositions can be made, some being commercially available. One grade was used for thermal insulation on the space shuttle. However the fibres are largely crystalline and there are no indications in the literature that glass fibres can also be made by sol-extrusion. Ray (1978) has described the development of similar materials, some of which have also been developed commercially.

A number of workers (Sakka 1988a; La Course 1988) have made silica fibres by drawing from the surface of certain sols when gel formation had proceeded to such an extent that the viscosity was in the range 1–10 Pa s. Some 20 μ diameter fibres, fired at 800°C, are reported to have strengths of 600 MPa.

Techniques for making microporous filters and membranes have been described (Kaiser and Schmidt 1984; Klein 1988b). Klein gives an account of the early stage in the development of a process, which involves pouring the sol to form a thin layer onto an inert liquid (tetrabromoethane) chosen for its high density and high surface tension. After gelling, the sheet is lifted off the liquid bath and then dried.

Considerable progress has been made in the manufacture of hollow glass microspheres, which are now made in production quantities by a sol–gel process. (Downs 1988). They are used in inertial confinement fusion research but also as fillers in various types of composite material. The processes used to make the confinement shells are described in considerable detail. Alkoxide gels are prepared as described above. After drying, they are comminuted and sized. The final step involves forming the microspheres by dropping the granules through a 4 m high drop tower in which the temperature is in the range 1500–1650°C. The paper gives a clear indication of the considerable effort required to make a practically useful product by a sol–gel process and a very valuable account of the many factors which have to be taken into account.

Several applications have been proposed for organically modified silicates (Philipp and Schmidt 1984; Schmidt 1984). Soft materials of high oxygen permeability and good wettability have been made by reaction of an epoxysilane with various titanium alkoxides. The material has promise for use as a hard contact lens material. The account of this development provides a good illustration of the scope within the class of sol–gel materials for tailoring the composition to obtain a combination of properties which is suitable for a given application.

By incorporating phenyl groups. thermoplastic materials have been made which are being developed as hot melt adhesives to form moisture-resistant seals between aluminium foil and glass containers. Abrasion – resistant coatings for polymers have been made, although products for this application have been commercially available for several years (Dow-Corning 1974). In the commercial product, colloidal silica is incorporated in a liquid phase made by the partial hydrolysis and condensation of an alkoxysilane. After coating, the material is cured at temperatures between 75 and 125°C. Coating thicknesses can be controlled between 0.5 and 20 μm.

Although outside the scope of this book, it is worth mentioning the applications of aerogels – extremely porous, low density solids which are made by the sol–gel process. The gel is dried supercritically. High quality transparent tiles have been made for use in Cerenkov counters for nuclear physics research. Fricke (1988) describes their manufacture and discusses their possible application as thermal insulating materials. They have considerable promise in this field. Used to fill the air gap in a double glazed window, the heat loss through the window would be reduced by a factor of two or three. However it would be necessary to evacuate the space between the panes to obtain the full benefit of using such a material. That would require a considerable change in the design and manufacture of the units.

The reader is referred to the monograph by Jones (1990) and papers by Mackenzie (1988) and Ulrich (1988) for far more comprehensive reviews of possible applications.

Chapter 10

Optical Applications

10.1. Introduction

Most applications of glass depend to some extent on its optical properties, its use in windows being a very obvious example. In this chapter a number of applications have been brought together in which the optical properties of glass are of primary importance and where the precise control of those properties has been essential for the development of new applications and even of new industries.

It is convenient to divide the chapter into two main sections, the first dealing with applications which depend on the refraction properties and the second with those which depend on the transmission properties of the material. Some applications do not lend themselves to such a simple classification. Thus the use of glass fibres for optical communications depends on both the refraction and transmission properties. This application is included in the first part of the chapter because the waveguide properties of the fibre, which control the propagation of light along it, depend on the variation of refractive index across its diameter. This aspect seems to be more closely related to the fundamental physics of the application than the problem of glass transmission, which is only one of the factors affecting energy loss.

Laser action depends on the fluorescence of ions dissolved in the host material (i.e. glass, so far as this book is concerned). This involves a mechanism which is different from those responsible for either refraction or absorption. Laser glasses are described in a short final section.

10.2. Lens systems

Optical glasses and the lenses made from them have been of the greatest importance for the development of many branches of pure and applied science e.g. medicine, biology, metallurgy and astronomy.

The range of lenses now available is very wide. At one extreme are spectacle lenses, made by the million, and at the other satellite camera and TV zoom camera lenses (Fig. 106) containing many components and using several types of glass.

The design, manufacture and testing of lenses is a well developed and complex technology in its own right and the last two aspects are dealt with in detail in excellent and interesting texts by Horne (1978, 1980, 1983) The standard method for producing the refracting surfaces is by mechanical grinding and polishing and this is likely to continue to be the practice for larger lenses. However an interesting recent innovation has been the development of techniques for making small high-precision aspheric lenses by direct moulding of the hot glass, so that no grinding and polishing is required. According to Garfinkel (1986), lenses about 1 cm in diameter can be moulded so that the lens replicates the mould to within 0.1λ.

10.3. Fibre-optic communications

10.3.1. Early developments

In the early 1960s a number of laboratories throughout the world were carrying out research on high capacity communication systems working at very short wavelength microwave and optical frequencies (Gambling 1986). In 1966 Kao and Hockham showed that optical communication systems could probably be developed based on cladded glass fibres, provided that the glasses could be made substantially free from transition metal ion impurities which absorb light in the visible and the near infrared. Currently available high quality optical glasses had absorption losses of the order of 1000 dB/km and it was considered that it would be necessary to reduce this figure by a factor of 50 for the proposed system to be viable.

Shortly afterwards Jones and Kao (1969) showed that a commercially available high purity silica glass, made by flame oxidation of $SiCl_4$, had a loss of only 10 dB/km, well below the required limit. The process had been known for some years and had been used to make high purity crucibles for melting semiconductor-grade silicon. Moreover

110

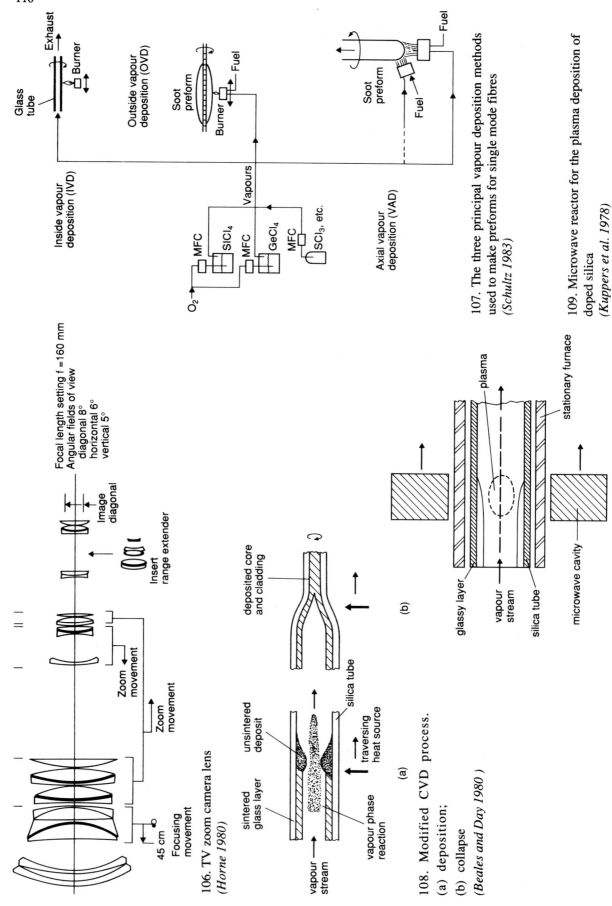

Inside vapour deposition (IVD)

Glass tube

Burner

Exhaust

Outside vapour deposition (OVD)

Soot preform

Burner

Fuel

Vapours

O₂

MFC

SICl₄

MFC

GeCl₄

MFC

SCl₃, etc.

Axial vapour deposition (VAD)

Soot preform

Fuel

Fuel

107. The three principal vapour deposition methods used to make preforms for single mode fibres (*Schultz 1983*)

109. Microwave reactor for the plasma deposition of doped silica (*Kuppers et al. 1978*)

Focal length setting f =160 mm
Angular fields of view
diagonal 8°
horizontal 6°
vertical 5°

Image diagonal

Insert range extender

Zoom movement

Zoom movement

Focusing movement

45 cm

106. TV zoom camera lens (*Horne 1980*)

deposited core and cladding

silica tube

(b)

unsintered deposit

silica tube

traversing heat source

sintered glass layer

vapour phase reaction

vapour stream

(a)

108. Modified CVD process.
(a) deposition;
(b) collapse
(*Beales and Day 1980*)

plasma

stationary furnace

glassy layer

vapour stream

silica tube

microwave cavity

Schultz and Smyth (1970) had used the same process to make silica glass doped with TiO_2, a material having a higher refractive index than the pure material. It was not surprising therefore that existing knowledge and techniques were quickly brought together or that the first silica fibres made by vapour-phase techniques and with a loss less than 10 dB/km were soon announced by Kapron, Keck and Maurer (1970).

Subsequently, considerable developments have taken place including new methods for producing the fibre preforms, further reductions in fibre losses, and the development of new lasers and detectors. A new industry now exists, based on the properties of glass and quickly brought into existence by the skills and knowledge of people who knew how to effectively manipulate those properties. Several good reviews exist e.g. Gambling (1986); Maurer (1976, 1977); Gossink (1977); Beales and Day (1980); Schultz (1983), and Geittner and Lydtin (1989); the last four being primarily concerned with the science and technology of preform and fibre manufacture.

10.3.2. Manufacture of preforms

There is still some interest in making high-purity, multi-component glasses from high purity but otherwise conventional batch materials for making into communication fibres, but most interest centres on silica-based fibres in which the preforms are made by a vapour phase process. The raw materials $SiCl_4$, $GeCl_4$, BBr_3 and $POCl_3$ are readily purified by distillation to remove light-absorbing impurities such as $FeCl_3$ and $NiCl_2$ (Gossink 1977). Fig. 107 shows in a single diagram the three principal methods which have been used to deposit doped silica inside or outside a silica tube, or on the end of a silica boule. The composition of the layer being deposited can be varied and the required radial refractive index gradient obtained by controlling the carrier gas flow through each of the flasks containing the liquid chlorides. The radial refractive index gradient in the fibre is of primary importance in determining the propagation of light along it.

Probably the most widely used deposition process, because of its speed, is the modified CVD process (MCVD) shown in Fig. 108. In the classical CVD process, the layer is built up directly from the gas phase onto the substrate (Chapter 7). In MCVD, oxidation of the halides occurs in the gas phase and the oxide smoke which forms inside the tube settles onto the wall and is subsequently sintered to form a dense material.

Another rapid deposition process is plasma-assisted CVD (Kuppers et al. 1978) illustrated in Fig. 109. The process operates at a reduced pressure and a plasma is produced by microwave energy within the tube. This activates the oxidation process and the material is deposited directly on the inside wall of the tube as in classical CVD.

10.3.3. Loss mechanisms

It is obviously important to keep the content of absorbing impurities to a minimum, especially those which are strong absorbers at or near the operating wavelength. There are also losses intrinsic to the pure material All the losses are wavelength dependent, so their relative importance depends on the choice of the wavelength at which the system is to operate – a choice limited by what lasers and detectors are available. The most important losses are:

    a. The fundamental absorption bands of the pure glass in the ultra-violet and especially in the infrared.

    b. Absorption due to chemically combined water, present as –OH groups in the glass. This can be reduced by ensuring that the gas streams are as dry as possible and, by excluding the ambient atmosphere during preform and fibre manufacture (Nassau and Shiever 1975). In silica glass the –OH group has its fundamental absorption at about $2.8\ \mu$, but even the second overtone at $0.9\ \mu m$ is strong enough to cause significant losses.

    c. Absorption due to ions of transition metals, many of which have absorption bands in the visible and near infrared. They originate from the raw materials or are picked up during processing.

    d. Absorption due to atomic defects arising from irradiation damage or from the non-stoichiometry of some glass component. Atomic defects with associated absorption bands may also be produced by the process of fibre drawing (Griscom 1978b, 1985).

    e. Scattering losses intrinsic to the composition of the glass.

We now consider some of these losses in more detail.

All the silicate glasses, including vitreous silica, are intrinsically highly transparent materials in the visible and near infrared. However silica glass has a strong absorption due to electron transitions in the near UV, so that even very pure material has a low transmission at wavelengths less than 200 nm, even in thicknesses as small as 1 mm. In the infrared, there is a strong fundamental absorption at $9.2\ \mu m$ associated with the stretching vibration of the Si–O bond. The first overtone of this band is also strong. Consequently millimetre thicknesses of silicate glasses are practically opaque at wavelengths greater than $4\ \mu m$. Directly related to these strong absorptions in the ultra-violet and the infrared are variations of the refractive index of the glass with wavelength as shown in Fig. 110.

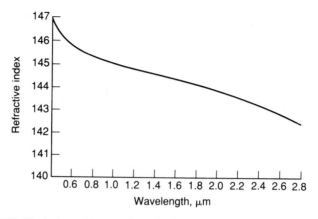

110. Variation with wavelength of the refractive index of silica
(Maurer 1977)

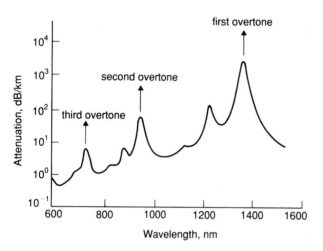

111. Absorption spectrum of hydroxyl groups in silica glass
(Keck et al.1973)

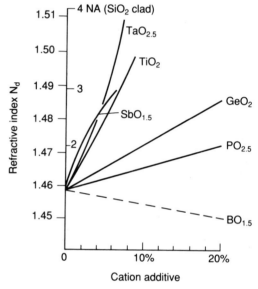

112. Effects of additives on the refractive index of silica glass
(Schultz 1983)

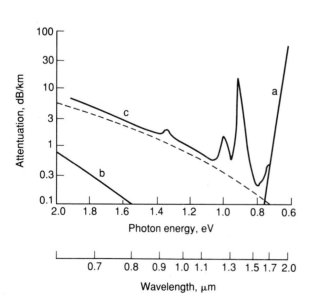

113. Measured loss compared with scattering and absorption losses

a. Infrared absorption

b. Ultraviolet absorption

c. Fibre loss

—— Rayleigh scattering

(Miya et al. 1979)

It can be shown that the fundamental absorption loss of pure, defect-free silica is at a minimum at the inflection point on this curve, i.e. at about 1.2 μm and that working at this point minimises pulse broadening. Initially the lasers and detectors available operated at 0.85 μm. The development of corresponding devices operating at a wavelength close to the optimum has resulted in a significant improvement in overall performance.

The absorption bands due to –OH groups in silica glass are shown in Fig. 111. The strong fundamental band due to the stretching vibration is at 2.8 μm with overtones at 1.4, 0.9 and 0.7 μm. Thus the overtone vibrations lie close to the operating wavelength, hence the importance of reducing the –OH content as far as possible.

One of the first doping materials to be used was TiO_2 which has the effect of increasing the refractive index. However this gave problems because of its tendency to dissociate to form Ti^{3+} which has a strong absorption in the UV. Dopants commonly used are GeO_2, P_2O_5, B_2O_3 and fluorine, the last two reducing the refractive index. Fig. 112 shows the effect of the oxide dopants on the refractive index.

It is well known that glasses discolour to a greater or lesser extent when exposed to ionising radiation or to UV. In vitreous silica, the strong radiation-induced absorption bands are in the UV, i.e. well away from the operating wavelength. Nevertheless care is needed when selecting compositions for fibres which may be exposed to high levels of radiation.

The final loss mechanism to be discussed is scattering. All transparent materials scatter light to some extent, even carefully purified and filtered liquids such as water and benzene. The scattering is caused by fluctuations in the refractive index of the material on a scale smaller than the wavelength of light. In glasses, the fluctuations present in the melt are frozen in on cooling at a temperature close to the transformation temperature. The glasses used in the silica communication fibres behave as Rayleigh scatterers, the intensity of scattering being inversely proportional to the fourth power of the wavelength. Consequently scattering losses are reduced by moving the operating wavelength further into the infrared.

The loss spectrum of a typical low-loss silica fibre is shown in Fig. 113. Apart from the overtones of the –OH absorption, the losses in the near infrared are only slightly greater than the theoretical limit set by the sum of the Rayleigh scattering and the tail of the SiO_2 absorption.

10.3.4. Fibre strength

The fibres are required to withstand high tensile stresses when they are assembled into cables and subsequently when the cables are laid. It is essential to apply a damage-resistant coating onto the fibre surface immediately the glass is cool enough to do so and before the fibre is wound onto the drum of the drawing machine. It is also necessary to prevent particles of dust becoming attached to the glass surface before it has been protected. A number of polymer coatings have been used, including thermally cured silicone resins and UV cured epoxyacrylates (Lawson 1984; Ta-Sheng Wei 1986, 1988; Clarkin et al. 1988)

A great deal of work has been done, applying the methods of flaw statistics to the analysis of fibre strength results so as to make it possible to predict the probability of fracture of kilometre lengths of fibre from measurements made in the laboratory on relatively short lengths. Fig. 114 shows the effect of increasing thicknesses of polymer coating on the strength distributions of 1 m long fibres (Huber and Gultmann 1980). Note that the thicker coatings not only increase the mean strength, they also greatly reduce the spread.

10.3.5. Heavy metal fluoride glasses

The accidental discovery by Poulain in 1974 of previously unsuspected but extensive glass forming systems based on the fluorides ZrF_4, ThF_4 and HfF_4 opened up the possibility of developing communication fibres with losses ten to a hundred times less than those of the silica fibres, i.e. in the range 10^{-2} to 10^{-3} dB/km at 2.5 to 3.5 μm. They have scattering losses which need be no higher at a given wavelength than those of silica glass. Since the fundamental IR absorption bands occur at longer wavelengths than in silica, it is possible to contemplate shifting the operating wavelength further into the infrared and so exploit the rapid decrease in scattering loss with increasing wavelength (Tran et al. 1984; Drexhage 1985).

Fig. 115 compares the IR transmission curves of two of these new glasses with that of fused silica.

There are many problems to be overcome before the technology of processing the fluoride glasses is as well developed as that for processing high purity silica. The viscosity–temperature curves of the new glasses are very steep, so fibre drawing will always be difficult. Also only recently have CVD techniques been developed (Fujiora et al. 1989), so most material has been made by crucible melting, with the associated greater risks of contamination. Inclusions are encountered in the melts, which may arise from impurities in the raw materials. Nevertheless it is clear from the loss measurements in Fig. 116 that encouraging progress has been made. The current state of the art, summarised by France et al. (1988), is that fibres approaching 1 km can be made with minimum losses approaching those of silica fibre.

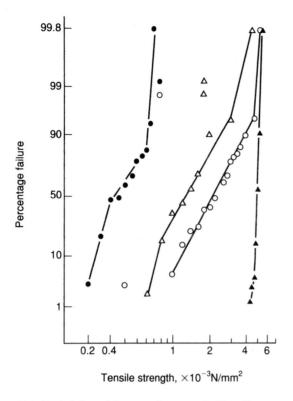

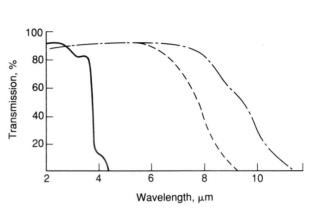

115. IR transmission curves of fused silica and two heavy metal fluoride glasses

——————— Fused silica

---------------- 57 HfF$_4$. 36 BaF$_2$. 3 LaF$_3$. 4 AlF3

-·—·—·—·— 17.5 BaF$_2$. 26.5 ZnF$_2$. 26 YbF$_3$. 30 ThF$_4$

(Tran et al. 1984 © IEEE)

114. Probabilty of fracture for coated silica fibres

○ 7 μm coating   Δ 13 μm coating
● 26 μm coating   ▲ 55 μm coating

(Huber and Gultmann 1980)

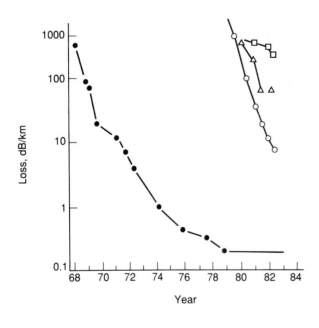

116. Historical record of the reduction in losses of various types of glass fibre

● silica

○ fluoride

Δ chalcogenide

□ Crystal

(Tran et al. 1984 © IEEE)

10.4. Signal glasses

The colours of the various types of transport signals must be carefully specified to ensure that they are easily distinguishable from one another. Pressed coloured glass lenses are widely used for this application. The signal colours most widely used are red, yellow, amber and green. White is also used in signals and its 'colour' has to be specified. Blue is used only when the illumination is high, partly because this colour is more scattered in foggy or smoky conditions and partly because the eye, especially for elderly people, is less sensitive to blue (Wright 1984).

The signal glass colours are specified in terms of the chromaticity of the light they transmit (e.g. by BS 1376: 1974). Fig. 117 shows the permitted chromaticity ranges. For each colour there are three classes A,B and C, class C being the most restrictive and therefore that used in especially critical applications. The British and other national standards also specify the minimum transmittance of each type of signal glass.

10.5. Glasses for infrared optics (Savage 1985, Worrall 1968)

10.5.1. General requirements

There are many applications for optical systems working in the infrared, e.g. for thermal surveillance and for heat-seeking missiles. The choice of operating wavelength depends on a number of factors, e.g. on the spectral sensitivity of the radiation detector, but two major factors which determine the designer's choice are the temperature of the target (which determines the operating wavelength) and the transmission of the atmosphere at this wavelength.

The presence of CO_2 and H_2O in the atmosphere results in three main transmission windows: 0.75–2.5 μm (near infrared), 3.0–5.0 μm (mid infrared) and 7.5–14.0 μm (far infrared). The last two are the windows of particular interest here. The mid infrared window is relevant for targets at relatively high temperatures ($> 400°C$) and the far infrared for targets at room temperature.

The wavelength regions of useful transmission for the oxide and chalcogenide glasses are limited at short wavelengths by electron transitions across the band gap and by lattice vibrations at the long wavelength cut-off. The latter has the greater importance in determining which type of glass is to be used in a particular IR system.

For glass thicknesses normally used (a few millimetres), the long wavelength cut off occurs at the first harmonic of the fundamental vibration frequency. For silicate glasses, this lies in the 4–5 μm region. Consequently these glasses are of little value, even for systems working in the mid infrared. Increasing the mass of the atoms which constitute the material and replacing the Si–O bonds by weaker bonds reduces the fundamental vibration frequency and displaces the long wavelength cut off to higher values.

10.5.2. Aluminate glasses

Several types of oxide glass have cut-off wavelengths beyond 5 μm (the aluminate, gallate and tellurite glasses), but by far the most important of these are the aluminate glasses (Worrall 1968).

The infrared transmission spectra of the oxide and chalcogenide glasses are markedly affected by impurities and it is necessary to remove them as far as possible. For the aluminate glasses, the important impurity is water dissolved in the glass and this can easily be removed by remelting the cullet *in vacuo*, which removes the –OH band in the 2.9–3.2 μm region (Fig. 118). Components weighing up to several kilograms can be made by casting and pressing but, because of the steep viscosity–temperature curve and ready devitrification at temperatures just above the annealing temperature, flame working of the glass is not possible. If seals are required to other materials, solder glasses may be used. The thermal expansion coefficient is approximately $95 \times 10^{-7}/°C$ i.e. about the same as that of a commercial soda–lime–silica glass, but the annealing temperature is much higher (800°C).

Worrall states that the mechanical properties of the glasses are good and (most surprisingly) that the strength is far less affected by static fatigue than that of flat glass, in spite of the fact that the aluminate glasses are appreciably attacked by humid atmospheres.

A detailed account of the melting and forming technology of the UK grade of aluminate glass has been given by Davy (1978).

10.5.3. Chalcogenide glasses

To obtain useful transmission in the far infrared region, it is necessary to use compositions containing elements in the chalcogen group (S,Se,Te). Kreidl (1983) has recently surveyed the literature on the formation and structure of glasses in these systems. Regions of glass formation are extensive, especially in ternary systems, but systematic research, especially by Savage, Hilton and their respective collaborators, has shown that there is only a limited number of compositions which have a satisfactory combination of mechanical properties and infrared transmission (Savage and Nielsen 1965; Savage *et al.* 1980; Savage 1985).

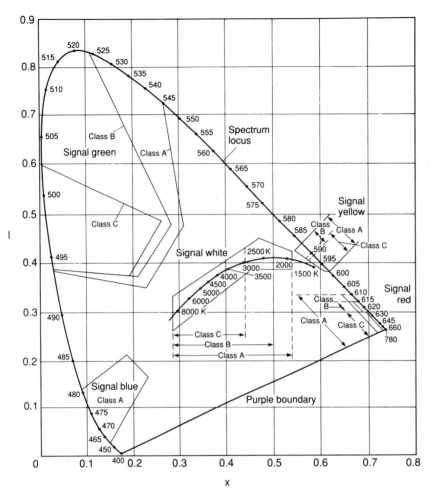

117. Chromaticity limits for signal colours
(BS 1376: 1974)

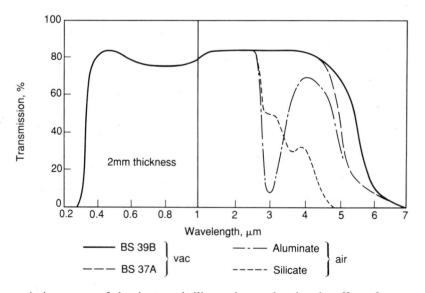

118. Transmission curves of aluminate and silicate glasses showing the effect of vacuum melting
(Worrall 1968)

Making the glasses and the elimination of impurities is difficult. One is dealing with elements and compounds which readily oxidise; some are volatile whilst others are toxic. It is therefore necessary to melt in sealed systems.

Sulphide glasses

The first chalcogenide glass to become commercially available was near-stoichiometric arsenic trisulphide. This is satisfactory for mid-infrared applications. After rough melting in a sealed steel or silica vessel, the material was purified by distillation. The vapour was condensed at a low enough temperature to maintain the As_2S_3 in the liquid state but high enough not to condense As_2O_3 impurities. Most of the oxide impurity was thus swept out of the still. The condensate was subsequently stirred. The material was produced in batch operations, each batch amounting to 3–5 Kg.

Composition control and homogenisation are not easy, mainly because the composition of the condensing vapour is not the same as that of the condensed material, hence the need for stirring. Worrall (1968) notes that because of the high refractive index of the glass, the reflection loss is high (17%). This can be greatly reduced by applying a polymer film to act as an anti-reflection coating.

Savage and Nielsen (1965) describe the development of several ternary Ge–As–S glasses, which have excellent infrared transmission from 1 to 11.5 μm. They have lower expansion coefficients and higher T_g temperatures than the binary glasses. One of these compositions was selected for commercial development. The materials were freed from oxide impurities by distillation in hydrogen and were melted in thick-walled silica tubes.

Fig. 119 shows the infrared transmission spectra of a number of ternary chalcogenide glasses and the marked improvement which results from the removal of the oxide impurities.

Table 17 gives the physical properties of a number of the glasses.

Table 17
Thermal properties of Ge–As–S glasses
(Savage 1985)

Composition	T_g °C	Thermal expansion coefficient. $(\times 10^{-7}/°C)$
As40. S60	165	261
Ge15. As25. S60	—	194
Ge25. As15. S60	425	128
Ge30 As15 S55	400	96
Ge40 As15 S45	—	77

Selenide glasses

As with the sulphide glasses, commercial development of selenide glasses has concentrated on that limited range of ternary compositions, which is considered to have the best combination of properties. They are glasses in the systems Ge–Sb–Se and Ge–As–Se. The spectral transmittance of one of these compositions is shown in Fig. 119b. The glasses are suitable for use in the far infrared window and some compositions of this type have been used to correct chromatic aberrations in infrared transmitting germanium lenses. The preferred composition range is Ge30, As or Sb 10–20, rest Se. Commercially available compositions are (in the UK) Ge30 As15 Se55 and (in the USA) Ge28 Sb12 Se60.

A considerable amount of work has been done to reduce oxide impurities, the most effective method being to vacuum bake the silica melting tube and the raw materials before melting. It has proved difficult to remove the last traces of impurity, which are believed to originate from volatilisation of the silica melting ampoule during sealing off. Although the experimental melts were made on a small scale, the process was scaled up to produce 1.5 Kg melts.

A very interesting CVD method for making high purity selenide glasses has been described in a patent by Katsuyama et al. (1982). The glass is deposited on the inner wall of a heated lead glass tube by passing through it a stream of argon carrying a mixture of $GeCl_4$, $SbCl_5$ and Se_2Cl_2 vapours. The tube was then collapsed and drawn to a fibre.

Selenide glass fibres have been made into fibre optic faceplates and have also been used to transmit CO_2 laser radiation for welding and cutting operations. Bornstein et al. (1985) have described methods for producing selenide glass rods and their drawing down into fibres. They found that an effective method for reducing the oxide impurity content was to add urea to the materials in the melting tube. This decomposes to form CO, NH_3, N_2 and H_2, gases which reduce any oxides present to the elements. The same method has been used by Lezal and Srb (1977). An interesting account of the development of infrared optical glasses in Czechoslovakia has been given by Vasko and Wachtl (1977).

10.6. Laser glasses (Snitzer 1966, 1973; Young 1969; Patek 1970; Neuroth 1987)

The first report of laser action using a glass host material, was by Snitzer in 1961 using a potassium barium silicate glass containing 2-wt% neodymium oxide. Since then much research has been carried out to determine the effects of glass composition on the parameters which affect the laser performance of glasses doped with rare earth ions. The Nd^{3+} doped glasses have been studied in most detail but the behaviour of other fluorescent rare earth ions has also been examined, e.g. Yb^{3+}, Ho^{3+}, and Er^{3+}. In addition to oxide host glasses (including tellurite and germanate as well as the more familiar silicate, borate and phosphate systems) the work has included a number of halide and oxyhalide systems (Weber 1982, 1983).

Wong and Angell (1976) give a simple explanation of laser action and of the advantages, for some applications, of using a glass rather than a crystalline host material. The fluorescent line width of Nd^{3+} in a glass is about thirty times greater than for the same ion dissolved in a crystalline material. This is a result of the greater perturbation of the relevant Nd^{3+} electron energy levels by the glass structure. As a consequence, the optical pumping power required for laser action is increased.

However the same feature is an advantage for pulsed operation, when the aim is to store a large amount of energy in the upper energy level before stimulated emission is allowed to take place. By using a neodymium glass laser in a Q-switched mode of operation and with mode locking, extremely short pulses (10^{-12} s) are produced. Thus even when only a modest amount of energy is stored in the laser, the power in the beam on discharge may be of the order of terawatts.

This feature is being exploited in a number of national facilities for fusion research, the most powerful of which is the Shiva facility at the Lawrence Livermore Laboratory (Speck et al. 1981; Hunt et al. 1981; Yamnaka et al. 1981; Ross et al. 1981). The laser energy is focused onto a glass microsphere containing the isotope mixture. Shiva is a 20 beam laser, the final glass component in each beam having a clear aperture of 20 cm diameter. The energy delivered is up to 10 kJ corresponding to a power of 30 terawatts.

The papers quoted give considerable detail on the technical facilities of each fusion installation but very little information about the glass composition and properties. It would appear that in most, if not all cases, it is a Nd^{3+}-doped 'phosphate' glass (probably an alkali or alkaline earth aluminophosphate).

For high power applications, it is necessary to take into account the fact that the refractive index of the glass increases with increasing beam intensity. The effect is not large – only a few ppm. Yet this is sufficient to result in self de-focusing and a significant loss of energy delivered onto the target. Thus it is necessary to select a glass with a low value of nonlinear refractive index coefficient, n_2.

Laser-induced damage is also a problem. The glasses are made in the same way as other optical glasses i.e. they are melted in platinum alloy crucibles and homogenised with stirrers coated with the same material (Jiang et al. 1986). Any small particles of platinum left in the glass may act as damage sites. The same is true of particles of foreign material on the glass surface. Thus the glass must be of extremely high quality.

According to Jiang et al. (loc cit), phosphate glasses were chosen because of their large stimulated emission cross section, low non-linear refractive index and good physical, chemical and mechanical properties. Their account of the manufacturing method used is of considerable general interest.

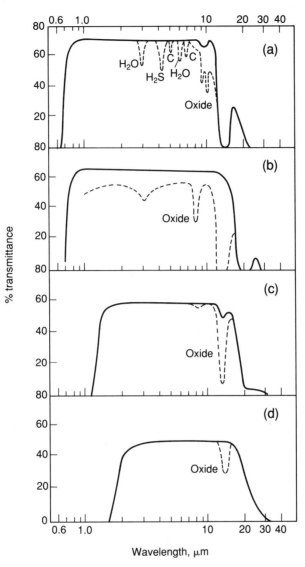

119. The infrared transmission curves of some chalcogenide glasses

(a). sulphide glass. Ge 30 As 20 S 50. 1.86 mm thick

(b). selenide glass. Ge 34 As 8 Se 58 1.80 mm thick

(c). selenide-telluride glass. Ge 30 As 13 Se 27 Te 30. 2.3 mm thick

(d). telluride glass. Ge 10 As 50 Te 40. 1.62 mm thick *(Savage 1985)*

E

Chapter 11

Applications of Electrically Conducting Glasses

11.1 Introduction

In well established electrical applications, e.g. those involving glass- to-metal seals, silicate glasses of conventional composition are normally used, a requirement being that they are sufficiently good electrical insulators at the working temperature.

In this chapter, we shall be concerned with materials at the opposite end of the range of conductivity values – with glasses of moderately high conductivities and their applications. The first section deals with glasses having abnormally high ionic conductivities (fast ion conductors) and their possible use as battery electrolytes. The second deals with oxide and chalcogenide semi-conducting glasses.

11.2. Glasses of high ionic conductivity for battery electrolytes

11.2.1. Glass compositions

Commercial oxide glasses have room temperature conductivities of the order of 10^{-16} S cm^{-1}. Corresponding values for some fast ion conducting (FIC) glasses are as low as 10^{-3} S cm^{-1}. Fig. 120 shows the variation of conductivity with temperature for one such glass, together with data for fused KNO_3. The $Ag_5I_4BO_3$ glass at T_g has a conductivity about the same as that of a typical molten salt.

Some FIC glasses are being developed as electrolyte materials to be used in batteries in reserve power supplies for computer memory boards. Others may be used in sodium–sulphur power storage batteries. In both applications a FIC glass is an alternative to a FIC polycrystalline (ceramic) material. There may be fabrication and processing advantages arising from the use of glass, but other factors have to be taken into account, especially the absence of reaction with electrode materials. Selection of the electrolyte material to be used in any application will depend on many factors.

The properties of many FIC glasses have been investigated. Most are oxide or sulphide glasses, 'doped' with high percentages of a silver halide or an alkali metal halide. An explanation of the high ionic conductivity which covers all compositions is not to be expected. However Ingram (1985) points out that, in solid electrolytes generally, the cation mobility increases with increasing polarisability of the anion sublattice. This is in accordance with the observation that the highest conductivities are obtained in the sulphide glasses doped with iodides. The ability to make glasses with a high concentration of alkali or silver cations is also an important factor.

Tables 18 and 19 give an indication of the unusual compositions of the glasses being studied (Minami 1985; Ravaine 1985).

Note from Table 19 the high concentration of alkali halides that can be introduced. All these compositions have relatively low T_g temperatures. Glasses of higher softening point are required for use in sodium-sulphur cells which operate at ca. 300°C. It has been suggested that an FIC glass of composition $Na_2O–2B_2O_3–0.16NaCl–0.2B_2O_3$ has a high enough T_g to be used in place of the current electrolyte material – β alumina. Although the conductivity of this glass is rather low, it may be possible to compensate for this by making the electrolyte in the form of a large number of thin walled tubes so as to give a large conducting area. This seems over optimistic. Commercial glasses of high silica content can certainly be drawn into fine tubing very easily, but this does not apply to low silica glasses or other unusual compositions. Another problem is that of developing a glass which has sufficient resistance to attack by sodium. It is well known from experience in the manufacture of low pressure sodium vapour discharge lamps that silica- containing glasses are readily attacked by sodium at temperatures of 300°C. Only aluminate and

Table 18 (Minami 1985)

Fast ion conducting glasses containing Ag^+ and Li^+ ions

Composition (mol%)	σ_{25} (Siemen/cm)
75 AgI. 25 Ag_2MoO_4	1.4×10^{-2}
85 AgI. 15 $Ag_4P_2O_7$	1.8×10^{-2}
60 AgI. 30 Ag_2O. 10 B_2O_3	8.5×10^{-3}
60 AgI. 40 Ag_2SeO_4	3.1×10^{-3}
33 AgI. 33 Ag_2O. 33 GeO_2	1.3×10^{-4}
50 AgBr. 25 Ag_2O. 25 B_2O_3	2.6×10^{-3}
40 AgCl. 30 Ag_2O. 30 B_2O_3	6.4×10^{-4}
44 LiI. 30 Li_2S. 26 B_2S_3	1.6×10^{-3}
45 LiI. 37 Li_2S. 18 P_2S_5	1.0×10^{-3}
50 Li_2S. 50 GeS_2	4.3×10^{-5}
17 LiCl. 25 Li_2O. 58 B_2O_3	1.9×10^{-7}

Table 19 (Ravaine 1985)

Conductivities and activation energies for glasses containing salt additives

Composition (mol)	σ_{25} (Siemen/cm)	E_0 (ev)
$LiPO_3$	2×10^{-9}	0.70
0.7 $LiPO_3$. 0.3 LiCl	1×10^{-7}	0.6
0.7 $LiPO_3$. 0.3 LiBr	3×10^{-7}	0.55
0.7 $LiPO_3$. 0.3 LiI	1×10^{-6}	0.52
0.33 P_2S_5. 0.66 Li_2S = A	1.1×10^{-4}	
0.29 A + 0.71 LiI	1.0×10^{-3}	0.31

Composition (mol)	σ_{200} (Siemen/cm)	E_0 (ev)
0.64 B_2O_3. 0.36 Li_2O = B	2×10^{-6}	0.72
0.54 B + 0.46 LiCl	2.5×10^{-3}	0.46
0.62 B + 0.38 Li_3SO_3	4×10^{-4}	

aluminoborate glasses are sufficiently resistant. Reference was made in Chapter 2 to the process for making tubing for sodium vapour lamps in which it is possible to make a composite tubing, combining the good workability of a silicate glass with the sodium resistance of an aluminoborate glass coating. It is difficult to see how a similar approach could be used to make tubing with a satisfactory combination of properties for this application.

Fig. 121 shows schematically the structure of a Na–S cell and Fig. 122 a specific design proposed by Levine (1980). The Levine cell contains several thousand hollow fibres made from FIC glass, each having an internal diameter of 50 μm and an external diameter of 80 μm.

11.2.2. Compatibility experiments and battery trials

Tuller and Barsoum (1985) and Gabano (1985) have summarised work designed to assess the long-term compatibility of glass electrolytes in contact with electrode materials.

Fig. 123 shows results of Velez *et al.* (1982) for the rate of growth of the reaction layer thickness on heating glasses of various Li_2O–B_2O_2 ratios in contact with lithium metal at 250°C. As for the reaction with other alkali metals, especially sodium, the rate decreases with increasing basicity of the glass (Elyard and Rawson 1962). Results were also obtained for glasses with alkali halide additions.

Gabano provides a considerable amount of information on the performance of test cells constructed as shown in Fig. 124. For these small cells (diameter 10.7 mm, thickness 1.2 mm, electrolyte thickness 0.5 mm) voltage-current curves were determined at 20° and 110°C. No serious compatibility problems were observed between the electrolyte and the cathode or anode materials. The work eventually led to a selection of electrolyte and electrode materials giving a cell of superior performance to that of commercially available designs. The paper is an interesting

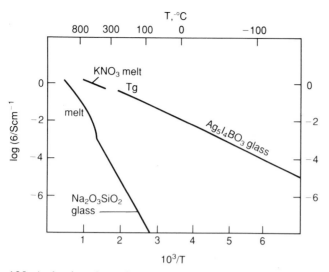

120. Arrhenius plots of conductivity for:
(1) KNO_3 melts.   (2) $Na_2O.3SiO_2$ melts and glasses)
$Ag_5I_4BO_3$ glasses
(Ingram 1985)

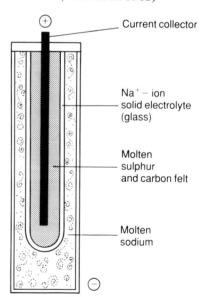

123. Effect of composition on reaction layer thickness
of glasses in the $Li_2O–B_2O_3$ system in contact with
molten Li at 250°C *(Vélez et al. 1982)*

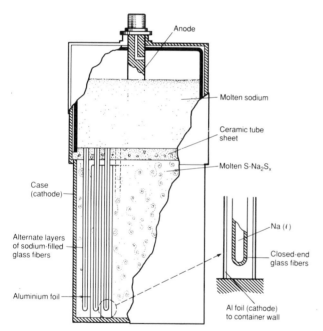

122. Schematic diagram of Dow cell
(Tuller and Barsoum 1985)

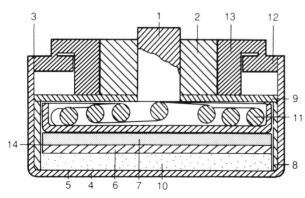

1. Central pin
2. Glass insulator
3. Laser welded seal
4. Stainless steel can
5. Cathode
6. Electrolyte
7. Anode
8. Insulating ring
9. Insulating disc
10. Stainless steel grid
11. Spring leader
12. Cover
13. Eyelet
14. Anode current collector

124. Schematic diagram of test cell
(Gabano 1985)

121. Schematic diagram of Na/S cell
(Ingram 1985)

and detailed account, rare in the literature, of the following through of a development, involving decisions on design and materials selection, to a promising conclusion.

11.3. Semiconducting glasses and their applications

The oxide and chalcogenide semiconducting glasses have room temperature conductivities which cover a wide range of values (see below). The oxide glasses contain high percentages of transition metal oxides, especially V_2O_5, and owe their conductivity to the presence of a mixture of valence states, e.g. V^{4+} and V^{5+}. Conductivity is due to the transfer of electrons between adjacent ions of different valency. Some chalcogenide semiconducting glasses are broadly similar in composition to those used in infrared optics (Chapter 10), but others differ considerably and the electrical properties of a wide range of compositions have been studied.

The discovery of chalcogenide and oxide semi conducting glasses (Kolomiets 1964; Baynton *et al.* 1957) at first attracted little interest. More recently a considerable amount of research has been carried out especially on the physics of electronic processes in these glasses. This has been summarized in a number of texts and review articles (Owen 1977, 1985; Mott and Davis 1979; Elliott 1983; Livage 1985). The understanding obtained has been of great value in developing applications for amorphous semiconductors in general, e.g. amorphous silicon, used in solar photovoltaic cells (LeComber 1985). (For a general review covering other applications, see Hajto and Owen 1984.)

A great deal of experimental and theoretical work has also been carried out on simple switching devices made using semiconducting oxide and chalcogenide glasses. The intensity of this activity has decreased considerably over the past five years, suggesting perhaps that more satisfactory devices can be made in other ways or from different materials.

11.3.1. Conductivity of chalcogenide glasses

Figs. 125 and 126 show respectively resistivity–temperature curves for a number of the simpler chalcogenide compositions and several glasses in the system V_2O_5–P_2O_5. An example of a more complex chalcogenide glass, used in the switching studies described below, is the composition Te39.As36.Si17.Ge7.P1. This material has a room temperature resistivity of 10^7 ohm cm.

The resistivity of the semiconducting oxide glasses can be varied over a considerable range by varying the melting conditions or by adding reducing agents to the batch. This is due to changes in the ratio of ions in the higher and lower valence states. Figs. 126 and 127 (Linsley, Owen and Hayatee 1970) illustrate the effect for several glasses in the system V_2O_5–P_2O_5.

11.3.2. Xerography

This is the most commercially important application of semiconducting glasses. Fig. 128 (Zallen 1983) shows the principle of operation of the familiar photocopier, which depends on the response to corona charging, followed by illumination, of a thin coating of a selenium rich glass. The coating is deposited onto the substrate drum by evaporation. Since a pure selenium glass would devitrify at the operating temperature, a small quantity (ca 1%) of Ge or As is added.

Fig. 128a shows the charging of the layer by the corona wire, producing a voltage gradient of about 10^5 V/cm across the film thickness. This step is carried out in the dark when the glass has a high resistance. Fig. 128b shows the charged layer being illuminated with an image of the document to be copied. In the bright regions of the image, electron/hole pairs are produced. Under the influence of the field, the electrons move to the surface where they neutralise locally the positive surface charges whilst the holes migrate to the substrate interface to locally neutralise the negative charges. The layer surface now carries a charge distribution which replicates the light intensity distribution which produced it.

The drum is next exposed to negatively charged toner particles which are attracted to the positively charged regions (Fig. 128c). These are minute particles of carbon each with a polymer coating. The particles are then transferred to a positively charged sheet of paper (Fig. 128d) which is finally heated to melt the polymer coating on the toner particles and so fix the image.

The advantage of using an amorphous semiconductor for the sensitive layer is that it can be made to have very uniform properties over a large area. Horne (1980) also gives a description of the process with more details of the optical and mechanical components.

11.3.3. Switching devices

Most of the work on semiconducting glass switches has used chalcogenide rather than oxide glasses, although similar effects have been observed using both types of material.

Two types of switching behaviour have been observed and both have been studied in great detail. They are illustrated in Fig. 129 (p. 125).

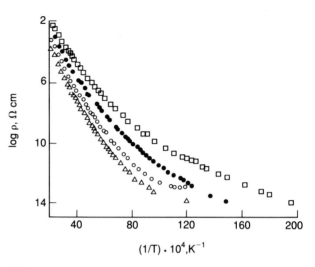

126. Temperature variation of conductivity for some
V_2O_5-P_2O_5 glasses
(Linsley et al. 1970)

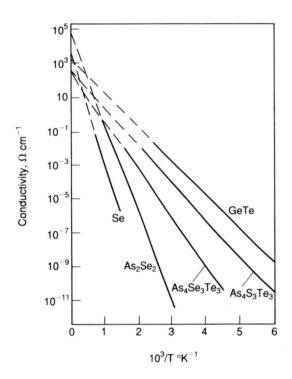

125. Temperature variation of conductivity for several
chalcogenide glasses
(Davies and Shaw 1970)

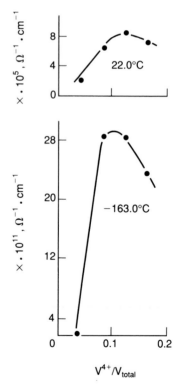

127. Effect of V^{4+}/V^5 ratio on conductivity of a
V_2O_5–P_2O_5 glass
(Linsley et al. 1970)

In a threshold switch, the characteristic has a stable and a metastable branch. The return to the high resistance branch occurs spontaneously when the current falls below I_h, the minimum holding current. A memory switch exhibits bistability: if a voltage greater than the threshold voltage V_{th} is applied for more than a millisecond or so, heating within the glass leads to the formation of a conducting crystalline bridge between the electrodes. The switch can be returned to the low conductivity state by passing a train of high current pulses, each having a sharp trailing edge. This results in melting of the bridge material which cools sufficiently rapidly for it to revert to its original glassy form.

Materials used for the two kinds of switch differ somewhat in composition: a typical threshold material is As30; Te48; Si12; Ge10 and a typical memory material is Te81; Ge14; Sb2 ; S3. Probably the most significant feature differentiating the two is that the memory materials lie closer to the edge of the glass-forming region in the respective system and are therefore more prone to devitrify.

Much work has been done in attempts to provide an understanding of the switching mechanisms, especially of threshold switching which presents more of a problem. Owen *et al.* (1979) conclude from a detailed experimental and theoretical investigation that threshold switching is primarily a thermal phenomenon whilst Adler *et al.* (1980) after an equally detailed study conclude that the effect is primarily electronic. It may be that the mechanism which is dominant depends on the glass used and the geometry of the device.

Very complex glass switch devices have been designed and made for particular applications, especially by Energy Conversion Devices Inc, a company directed by S.R.Ovshinsky, one of the original discoverers of the switching effect (Ovshinsky 1968). Descriptions of some of these have been given by Neale(1970); Nelson (1970); Fleming (1970); Maruyama *et al.* (1973); Chen *et al.* (1973) and Holmberg *et al.* (1979).

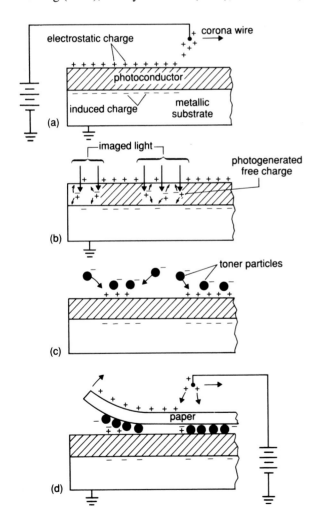

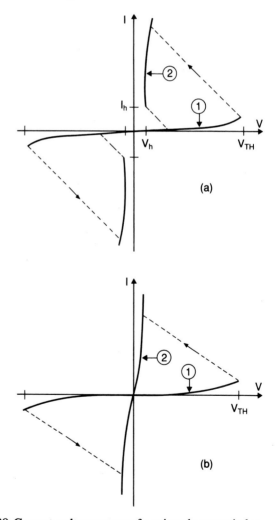

128. The use of amorphous semiconducting layers in xerography. (a) charging (b) exposure (c) development (d) transfer
(Zallen 1983)

129. Current-voltage curves of semiconductor switches
(a) Threshold switch
(b) Memory switch
(Owen et al. 1979)

Chapter 12

Metallic Glasses and their Applications[*]

12.1. Introduction

By cooling the melt extremely rapidly, many substances can be obtained as glasses which would crystallise if cooled at rates typical of those used in preparing the well known oxide glasses (Scherer and Schultz 1983). To be of any practical interest the material must be glassy throughout its thickness so the application of the rapid cooling approach is limited to the preparation of thin ribbon or fine wire and can be expected to be most effective when quenching materials of high thermal conductivity, i.e. metal alloys.

The first metallic glass had a composition near to the eutectic in the Au–Si system (Klement *et al.* 1960; Duwez *et al.* 1960) and was made by causing an atomised droplet of liquid alloy to impinge very rapidly on a highly conducting substrate. The estimated rate of cooling is $10^6 \,^\circ\text{C s}^{-1}$.

Although several quenching techniques have since been used to produce a wide range of metallic glass compositions, the most common is the melt spinning technique illustrated in Fig. 130. The melt is driven by gas pressure from the melting tube onto the surface of a rapidly rotating drum to form a continuous ribbon, typically 10–50 μm thick and 1–3 mm wide. Wider ribbon can be made by using a slot-shaped orifice. The figure indicates some of the process variables that affect the process and the properties of the material (Köster 1985).

There now exists an extensive literature on metallic glasses, resulting from research stimulated partly by a scientific interest in this new and intriguing class of materials and partly by the hope of exploiting some of their remarkable properties.

12.2. Metallic glass compositions

Cahn (1980) classifies the large number of metallic glasses into five categories (Table 20). He notes, as others have, that regions of glass formation are frequently associated with regions in which the liquidus temperature is low relative to the melting points of the constituent elements. A similar state of affairs exists in the oxide glass-forming systems.

Table 20

Classification of glass-forming alloy systems
(Cahn 1980)

Category	Representative systems	Typical composition range %
1. T^2 or noble metal + metalloid (m)	Au–Si, Pd–Si, Co–P, Fe–B, Fe–P–C, Fe–Ni–P–B, Mo–Ru–Si, Ni–B–Si	15–25 m.
2. T^1 metal + T^2 (or Cu)	Zr–Cu, Zr–Ni, Y–Cu, Ti–Ni Nb–Ni, Ta–Ni, Ta–Ir	30–65 Cu or T^2 or smaller range
3. A metal + B metal	Mg–Zn, Ca–Mg, Mg–Ga	Variable
4. T^1 metal + A metal	(Ti, Zr)–Be	20–60 Be
5. Actinide + T^1	U–V, U–Cr	20–40 T^1

Key to Table 20: A metal: Li, Mg groups. B metal: Cu, Zn, Al groups. T^1: early transitional metal (Sc, Ti, V groups). T^2: late transitional metal (Mn, Fe, Co, Ni).

[*]Davies 1976; Cahn, 1980; Elliott 1983; Van der Sande and Freed, 1983.

Table 21
Strength and tensile properties of various fibre-reinforcing materials
(Cahn 1980)

Material	Fracture strength σ_t GPa	Relative density ρ	$\sigma y/d$	Youngs' modulus E GPa	E/d
Glassy metals					
$Fe_{80}B_{20}0$	3.6	7.4	0.49	170	23
$Ti_{50}Be_{40}Zr_{10}$	2.3	4.1	0.56	105	26
$Ti_{60}Be_{35}Si_5$	2.5	3.9	0.64	ca.110	ca.28
$Cu_{50}Zr_{50}$	1.8	7.3	0.25	85	12
Other fibres					
Carbon fibre (PAN)	3.2	1.94	1.65	490	253
Glass fibre ($MgO–Al_2O_3$)	5.0	2.5	2.0	85	34
B filament (on W core)	2.5–4.5	2.6	1.0–1.7	380	146
SiC filament	3.5	2.6	1.35	200	77
Kevlar fibre	2.8	1.5	1.9	35	90
High C steel fibre	4.2	7.9	0.53	210	27

One material of composition $Fe_{40}Ni_{40}P_{14}B_6$ ('METAGLAS 2826'), commercially available from the US company Allied Chemical Corporation, has been widely used in property studies.

Davies (1976) has given a very detailed review of glass formation in metal alloy systems, and discusses various attempts which have been made to provide theoretical explanations for the regions of glass formation. As for the oxide systems, it appears that the structural explanations are not entirely convincing and that it is more helpful to interpret the information in terms of the kinetic theory of nucleation and crystal growth or some simple parameter which originates from that theory.

12.3. Mechanical properties

Metallic glasses have strengths approaching the theoretical fracture strength. Because they also exhibit some degree of ductility, they are tough and relatively insensitive to the types of damage that drastically reduce the strength of oxide glasses. The mechanisms of deformation have been studied in considerable detail. At room temperature, the plastic deformation is inhomogeneous and involves the formation of widely spaced shear bands. At higher temperatures near T_g the glass deforms by a homogeneous flow process. Table 21 compares the mechanical properties of several metallic glasses with those of other high strength materials (Cahn 1980).

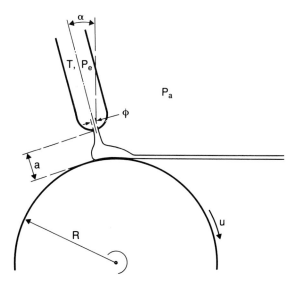

130. Schematic diagram of melt spinning process showing the parmeters that affect the process
(Köster 1985)

In commenting on this data, Cahn (1980) points out that the mechanical properties of metallic glasses are not so outstanding as to make them preferable to other high strength materials. For weight-sensitive applications, e.g. in aircraft, important parameters are σ_y/d (where σ_y is the yield stress) and the specific stiffness E/d. The values of these parameters for the metallic glasses are not especially high. However he considers that the main obstacle to the use of metallic glasses in structural applications is their cost. When this can be reduced, applications depending on the mechanical properties are likely to follow.

Davis (1985) seems to take a more pessimistic view. He points out that times to fracture by cyclic fatigue are rather low for metallic glasses. This is due to a lack of work hardening which, in steel, disperses slip initiated at surface or bulk inhomogeneities. He quotes results for the fatigue lives of reinforced polymers to illustrate the point. For a minimum lifetime of 10^6 cycles, a composite reinforced with metallic glass ribbon can be expected to withstand a peak stress of about 11% of the composite strength. The corresponding figure for a graphite fibre reinforced epoxy is about 70%.

12.4. Magnetic applications (Luborsky et al. 1978; Luborsky 1985)

The outlook for applications based on the magnetic properties of metallic glasses is more promising. The first ferromagnetic glass of composition $Fe_{83}P_{10}C_7$ was discovered by Duwez and Lin (1967). Many more were subsequently discovered and it appears that glassy alloys of this type are usually ferromagnetic if the corresponding crystalline alloy (minus the metalloid) is itself ferromagnetic. This development was met with some surprise, since all ferromagnetic materials then known were crystalline. However the significant factor in ferromagnetism is the spacing between the interacting spins, not whether or not these spins are arranged on a regular lattice.

Properties of some ferromagnetic glasses and comparable commercial magnetic materials are given in Table 22. The relatively low values of the coercive field are due to the absence of grain boundaries and inclusions which, in crystalline ferromagnetic materials, have the effect of pinning domain walls.

Table 22
Some properties or amorphous and crystalline ferromagnets
(Luborsky 1977)

Alloy	Coercive field (mOe)	Saturation induction at 298°K (kG)	Curie temp. °C	Magnetostriction ($\times 10^6$)
Amorphous				
$Fe_{80}B_{20}$	40	16.0	374	30
$Fe_{80}P_{16}B_1C_3$	50	14.9	292	30
$Fe_{40}Ni_{40}P_{14}B_6$	19	8.2	247	11
$Fe_{29}Ni_{49}P_{14}B_4Si_2$	11	4.2	382	5
$Fe_3Co_{72}P_{16}B_6Al_3$	15	6.3	260	0
The first four are commercial 'Metaglas' materials				
Crystalline				
$Ni_{80}Fe_{16}Mo_4$ 'Mumetal'	25	7.8	460	0
'Hymu 8000'	5	8.2	400	0
Oriented transformer steel	300–500	20.3	730	4

Luborsky (1985) gives an interesting account of the application of ferromagnetic glasses in the cores of transformers with ratings up to 25 kVA. Core losses are typically one third of those of a transformer made using the conventional steel. He calculates that if all the distribution transformers in the USA were replaced by transformers made using ferromagnetic glasses, the energy cost saving would be a quarter of a billion dollars per year. He describes a contract to his company (GE of America) to supply to a number of power utilities 1000 distribution transformers rated at 25 kVA. Several had already been built and had demonstrated satisfactory performance in service. The cores were constructed from 2.5 cm wide ribbon.

He also lists (Table 23) a number of small devices, available commercially, which use ferromagnetic glasses.

With some justification he writes 'I believe that the dawn of the industrial age of amorphous metals has already arrived as evidenced by the many applications of amorphous materials . . .'.

More specialised applications are being developed. Smith (1985) describes the successful use of ferromagnetic glass core saturable inductors as high capacity switching elements in pulse power sources for particle accelerators at the Sandia and Lawrence Livermore National Laboratories. The switching action depends on the large fall in reactance as the core saturates. By using a ladder arrangement of reactors and condensers, a voltage peak is compressed in time and increased in amplitude.

Gambino (1985) gives an interesting assessment of the possible applications in computer components of thin film amorphous ferromagnetic materials, produced either by sputtering or evaporation. Applications considered include bubble memories, magnetic and thermomagnetic recording and thin film heads. Device performance using amorphous materials is promising but marginal improvements in properties or advances in processing techniques are needed before these materials are likely to be widely used.

Table 23
List of commercially available magnetic devices using amorphous alloys
(Luborsky 1985)

Approx. date of introduction	Device
1980	Core in moving magnet recording cartridge.
1981	Audio, computer cassette heads. Transformer in dynamic microphones. Recording head for tape duplication. Transformer for moving coil recording head. Magnetic amplifier in switched mode power supply. Delay lines in data tablet. Multigap head for height gauge.
1982	Sensor in electronic '90' machine. Step-up transformer for moving coil cartridge. Recording head in electronic still camera. Cores for transductor chokes, high frequency small transformers, magnetic switches. Strip for magnetic recording heads, theft protection devices, magnetic shielding.
1983	Magnetic amplifier output regulator in switched mode power supply. Turns ratio transformer. 'C' cup cores on converter output. Current spike suppressors. Core in power supply modulator for laser. Digitising pad.

Chapter 13

Fibre Glass and its Applications

13.1 Introduction

Fibre glass is manufactured in various forms for a wide range of applications. Mohr and Rowe (1978) have described the manufacturing processes and the applications for all forms of fibre glass–glass wool for thermal or sound insulation and as a filter material, continuous filament fibre glass for use in composites. Loewenstein's book (1973) deals specifically with the manufacture and surface treatment of the continuous filament material and the SPI Handbook (Mohr *et al*. 1973) gives detailed information on composite processing methods and composite properties. Aubourg and Wolf (1986) give an account of the development of the various glass compositions. Manufacturers' publications give valuable information on the products available and on their applications.

13.2. Glass and mineral wool

Typical compositions are given in Table 24 (Mohr and Rowe).

Table 24
Typical compositions of glass and mineral wool materials

	Mineral	Glass insulation	High temperature fibre
SiO_2	50	63	50
Al_2O_3	10	}6	}40
Fe_2O_3	1		
CaO	25	7	6
MgO	14	3	4
Na_2O	–	14	–
K_2O	–	1	–
B_2O_3	–	6	–
F_2	–	0.7	–

The fibre is made by one of a number of manufacturing processes. In some, a large number of glass streams flowing from a multi-orifice platinum alloy bushing in the base of a feeder channel are entrained in high velocity steam jets or in the gases from a high velocity burner. In the widely used rotary or 'crown' process (Fig. 131), a single stream of glass from the furnace falls into a rapidly rotating nickel chromium or platinum alloy cylinder (Stokes1987). Centrifugal force drives the glass horizontally through holes in the cylinder sidewall, where the streams are attenuated by downwardly directed air jets or burners.

The fibres are sprayed with an organic binder (e.g. phenol–formaldehyde resin) before they reach the collecting belt. The mat is then passed through an oven to cure the binder. Several types of product are made, including loose wool for hand-packed thermal insulation, flexible rolls for thermal and acoustic insulation and rigid sheets, slabs and standard shapes, also for thermal and acoustic insulation. The binder plays a different role in each of these products. In the rigid products its function is obvious. In the mat, it makes handling easier and prevents loss of strength by self-abrasion of the fibres. It also ensures thickness recovery on unpacking by the user. The mat is compressed to about 25 % of its volume for packing and it is essential that the thickness recovers, otherwise some of the mat's insulating properties will be lost. The glass fibre must not release too much alkali under moist conditions or the bond may deteriorate and the recovery property will be lost. The binder may also contain additives for various functions, e.g. fire retardancy and water repellancy.

Table 25 compares the room temperature thermal resistances of fibre glass mat with that of other building materials.

Fig. 132 shows that the thermal conductivity increases significantly with temperature, due primarily to the increasing conductivity of the air trapped in the material.

Grades of fibreglass are available for furnace insulation made from refractory fibres (e.g. Al_2O_3–SiO_2). Maximum service temperatures are of the order of 1300°C. For insulation applications they are most cost-effective when used as back up to brick insulation.

An interesting and demanding use of fibreglass insulation is in the NASA space shuttle. The insulating cladding on the under surface of the shuttle has to withstand the re-entry conditions which involve surface temperatures up to 1280°C. The fibres are made from a standard fibreglass but subjected to a special acid treatment which increases its silica content from 58–97.5%. Mohr and Rowe describe the process used to make tiles from this material.

Table 25
Thermal Resistivities, R
(Mohr and Rowe 1978)

	R
Common brick (4" thick)	5.5
Expanded polystyrene (1")	27.8
Fibreglass roof insulation (15/16")	25.7
Fibreglass roll (3.5")	76.4

R = 1/(thermal conductivity in W/m °K)

13.3. Continuous filament fibre glass and composites

13.3.1. Compositions, processes and products

Continuous filament fibre glass is widely used to reinforce various polymers to make articles ranging from furniture to the hulls of small ships. The fibre can be woven into fabric which also has many uses – as a curtain material, in filters, for electrical insulation and for making composites, e.g. printed circuit boards. In recent years an alkali-resistant fibre glass has been developed for reinforcing cement and this has been widely used in building components.

Several compositions are in use, which are normally referred to by a letter designation, i.e. E glass, C glass, etc. The more important compositions are given in Table 26.

Table 26
Continuous filament fibre compositions

	E	C	AR	S	AR (Proctor 1985b)
SiO_2	54.3	64.6	60.9	65.0	62.0
Al_2O_3/Fe_2O_3	15.2	4.1	0.27	25.0	0.8
CaO	17.3	13.4	4.8	0.0	5.6
Na_2O	0.6	7.9	14.3	14.8	
K_2O		1.7	2.7		
B_2O_3	8.0	4.7			
BaO		0.9			
TiO_2			6.5		0.1
ZrO_2			10.2		16.7

E glass was originally developed for applications in the electrical industry. Being alkali-free it has a high electrical resistivity and good chemical durability in neutral and mildly acid environments. These qualities, together with its high strength when properly processed, have led to its wide use in polymer composites.

C glass has greater resistance to acid attack, hence it is the preferred fibre for use in lead–acid batteries. AR glass,

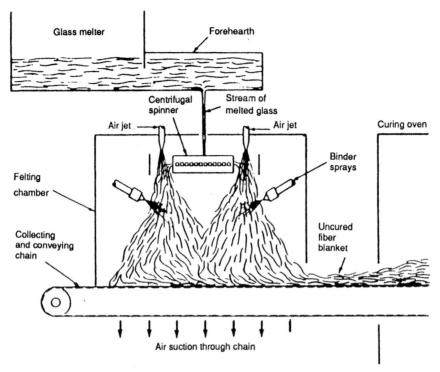

131. Rotary process for making glass wool
(Mohr and Rowe 1978)

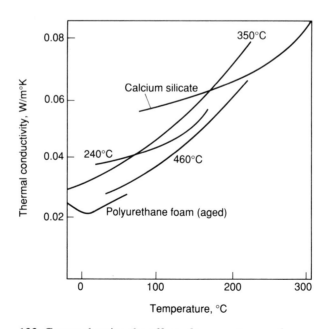

132. Curves showing the effect of temperature on the
thermal conductivity of three fibre glass pipe insulating materials having the given upper temperature limits. Curves for calcium silicate and polyurethane foam
are given for comparison
(Mohr and Rowe 1978)

more commonly known by its Pilkington trade name ' Cem-FIL' is an alkali resistant glass, an essential requirement for cement reinforcement and finally S glass is a high strength, high Young's modulus composition, occasionally used to make stiffer and stronger composites.

In the fibre forming process (Fig. 133), the glass flows through orifices in an electrically heated platinum alloy bushing. The traction from the winding drum stretches the fibres and pulls them through a mechanical device which twists them into a single strand of many fibres (a roving). Immediately above the strand former, the fibres are sized. This is partly to prevent damage by inter-fibre abrasion and partly to produce a surface which will give good bonding to the polymer used in the composite.

The figure shows fibre production by the direct melt process. Before this was perfected, the glass to be fiberised was remelted electrically in a small platinum lined melter, fed with glass marbles. These were produced in a conventional furnace, often in some other plant.

The fibre is supplied in various forms, especially for use in the many processes for making glass fibre reinforced plastics. Thus continuous fibre is used for filament winding and pultrusion systems (in which the continuous fibre is co-extruded with the resin). Continuous roving is also used when components are built up by co-spraying the fibre with the resin, the fibre being chopped into short lengths inside the spray gun. Reinforcing mat, consisting of a layer of short fibres lightly bonded together, is used in hand lay-up of products such as aircraft or boat components, tanks and ductwork. Short chopped strand is available for making moulding doughs and extruded components.

13.3.2. Glass fibre reinforced plastics

A considerable amount of information is available on the factors which determine the properties of these materials. A recent book by Ashbee (1989) deals in particular with the mechanical properties and with environmental degradation of the material. The Young's modulus of the fibre (68–76 GPa) is about ten times greater than that of polyester and epoxy resins. The modulus of the composite depends on the fibre volume fraction, η, and for the simplest case of a uniaxial arrangement of continuous fibres, the relationship is:

$$E_c = \eta \cdot E_f + (1-\eta) \cdot E_m$$

where E_c, E_f and E_m are respectively the Young's modulus of the composite, the fibre and the polymer matrix.

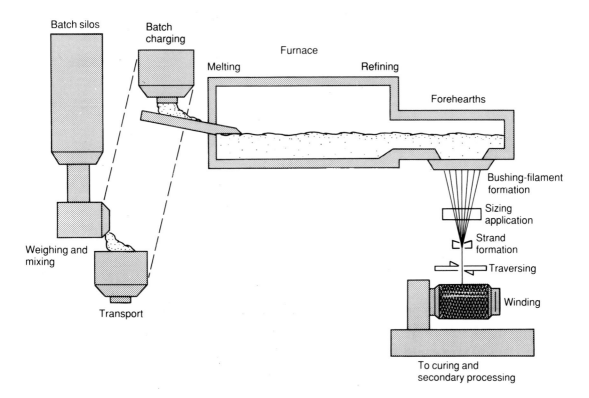

133. Schematic diagram of the direct melt process for production of continuous-filament fibre glass *(Mohr and Rowe 1978)*

However in most reinforced plastics, the fibres are only a few centimetres long and the composite modulus is therefore lower than is given by the equation. Typical values are in the range 7–14 GPa, whilst for filament wound products, in which the fibres are continuous, values are of the order 50 GPa. The modulus is anisotropic, especially when the fibres are laid within the resin in layers, the modulus normal to the fibres being then much less than in their plane.

The strength and toughness of the composite also depend in a complex way on relationships between the pattern of fibre orientation determined by the manufacturing process and the nature of the applied stress system. The fibre strength is very high (ca 2000MPa) but the glass has a fracture surface energy about 100 times less than that of the resin. The composite benefits only to a limited extent from the desirable properties of its components.

The strength of the adhesion between the resin and the glass is also important. In general this should be high, since the toughness of the composite is increased if large forces are required to pull fibres out of the resin when fracture occurs. On the other hand a weak interface between the resin and the fibre can have the valuable role of stopping any crack which approaches the interface.

For most composites the tensile strength is in the range 70–150 MPa, depending on the construction of the composite and the method of manufacture. However filament wound composites, in which the reinforcement is made by continuous winding of fibre onto a mandrel, are much stronger (500–1700 MPa).

Although the glass surface may appear to be protected against atmospheric influences by the resin, this is a misleading impression. The interfacial strength is reduced by water attack, the water having either diffused through the polymer or along the glass-polymer interface (Andrews et al. 1982). The resins used absorb a significant amount of water and expand in volume. For an epoxy resin, this amounts to a few percent at saturation. This stresses the fibre to an extent which varies along its length. Ashbee (1989) has described in detail a number of mechanisms by which water can enter the composite and eventually cause mechanical breakdown. The rate of degradation depends on the composition of the fibre, being slow for silica fibres and most rapid for an alkali- containing fibre, such as C glass.

The surface treatment given to the fibre before the composite is made is of great importance for its strength and durability. Sizes are applied containing some kind of coupling agent, which may be a silane or a chromium complex (Mettes 1969; Loewenstein 1973; Plueddemann 1981). The surface treatment must be capable of maintaining the strength of the composite throughout its service life. . Given correct selection of materials and processing methods, this apparently presents no serious problems, as is shown by the use of glass–polymer composites in marine applications. However, according to Proctor (1985a), if glass-reinforced plastics are to be used in wet conditions, long term design stresses are normally taken as a fraction (1/5 to 1/15) of the short term material strength.

13.4. Glass fibre reinforced cement

The early work on this material was carried out by Majumdar and his colleagues at the Building Research Establishment in the UK (Majumdar 1970). It was subsequently developed and marketed by Pilkington Brothers. Blackman has described its use in large building components (1974, 1979). The manufacturing method commonly used is to spray the cement slurry onto a suitable former; the AR glass fibre is fed to a chopper and feeder unit mounted on the gun assembly.

The novel feature which made this material possible was the development of a glass fibre which shows little loss of strength when exposed to the very aggressive alkaline environment of the setting cement paste (pH 11 to 12). Results quoted in Chapter 2 show that silicate glasses are severely attacked under these conditions by a mechanism which involves breakdown of the silica network. The attack is largely prevented in the AR glass by the relatively high percentage of ZrO_2 in the glass.

Larner et al. (1976) have carried out a detailed investigation of the chemical reactions between aqueous extracts of Portland cement and glass fibre of three compositions, including E glass and 'Cem-FIL'. Leaching of glass components was far less from 'Cem-FIL' than from the other two glasses. A hydrated calcium silicate layer, containing zirconia forms on the 'Cem-FIL' surface and it is suggested that this is responsible for the enhanced alkali resistance.

The ultimate commercial success of the development depends on obtaining the approval of national authorities throughout the world for GRC as a building material. Much laboratory and field work has been done to achieve this end (Proctor 1980, 1985a, 1985b).

Proctor has made considerable use of a simple strand-in-cement (SIC) test (Fig. 134) to investigate the factors affecting the strength of GRC composites and to make reasonably confident predictions of their useful service lives.

He shows (Fig. 135) that the strengths of the test specimens after immersion in water for various times and temperatures correlate very well with the flexure strength of GRC composites subjected to similar accelerated ageing conditions. Results from these tests also show that temperature is the main factor determining the loss in strength. Strength data obtained from measurements made in many parts of the world can be brought together on a single Arrhenius plot if an appropriate time shift is made, this depending only on the local annual mean temperature.

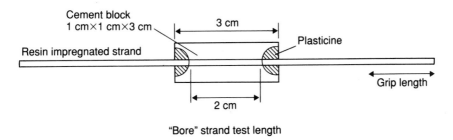

134. Strand-in-cement (SIC) test specimen
(Proctor 1985a)

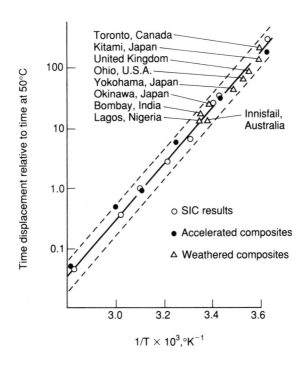

135. Normalised rates of strength loss in accelerated hot/wet ageing of SIC specimens and in natural weathering of GRC composites
(Proctor 1985a)

F

Chapter 14

Reactions with Aqueous Solutions:
Their prevention and some current applications

14.1. Introduction

All commercial glasses must have an adequate resistance to attack by the ambient atmosphere and in particular by the water vapour contained in it. A glass container must not react by more than a specified amount with its liquid contents. This is essential if the liquid is blood plasma or a pharmaceutical product. Such requirements have been familiar to the glass industry for many years and usually they are easily met.

However an application has recently arisen in which it is essential to limit the attack of glass by aqueous solutions to the absolute minimum. This is in the vitrification of fission product wastes from nuclear reactors. Although the glass will be stored in sealed, stainless-steel canisters and in what, at present, are dry sites, it is essential to choose a composition with the highest possible resistance to leaching by any water that may subsequently enter the site. This must now be the most thoroughly researched area of glass science and technology – for obvious reasons. Melting techniques and equipment must ensure the safe handling of the active wastes and extensive measurements of glass properties are needed to make possible reliable predictions of the behaviour of the material over many thousands of years.

Other recent developments and applications, are seeking to *promote* a reaction with an aqueous solution, provided that it is of the type required. When a glass or a glass–ceramic is used as a biological implant, the surface reaction with the surrounding body fluid will determine whether or not the implant will be successful, i.e. whether it will form a strong bond to adjacent healthy bone or other tissue.

Another recent application, based on controlled reactions, involves the use of glasses and vitreous enamels to release at a controlled rate essential trace elements to their environment. Such materials have been used, for example, for soil treatment, for the medication of animals suffering from the effects of trace element deficiencies and to kill harmful organisms in infested streams.

14.2. Pharmaceutical applications

The glass container industry and container users in the medical field rely on a number of international standards to determine the suitability of particular glasses or to test the containers themselves. When testing glass (ISO 719), the material is crushed to a powder of controlled particle size, which is exposed under controlled conditions to attack by boiling deionised water for 1 h. The alkali extracted is determined by titration. In a similar but more severe test (ISO 720), the reaction is carried out in an autoclave at 121 °C. In ISO 695 a *bulk* specimen of specified surface area is subjected to attack by a boiling solution of sodium carbonate and sodium hydroxide. The extent of reaction is measured by determining the loss in weight of the glass.

The sole purpose of the ISO standards in this field is to specify exactly the method of test and all relevant features of the equipment to be used. Although the standards list letter grade scales according to the range of values within which the results may lie, they do not specify the maximum extent of attack that is acceptable for any application.

The latter information may be found, for example, in European Pharmacopoeia (Volume I). This standard covers containers for injectable preparations (ampoules and small thick-walled containers). The containers are partly filled with deionised water and are heated in an autoclave for a specified time. The alkali extracted is determined by titration and limits are specified depending on the size of the container, i.e. on the area of glass exposed to attack.

14.3. Vitrification of high level nuclear waste

The literature on this subject is extensive, as one might expect. Three review articles (Laude 1985; Wicks 1985; Hayward 1988) give an insight into the many aspects involved.

Both Laude and Wicks deal with the storage of waste by vitrification in a sodium borosilicate glass, the preferred method at the present time. Laude describes the French plants at Marcoule and La Hague.

After calcining the waste solution (nitrates of the fission product elements) in a rotary kiln, the material is mixed with the glass frit which is then melted by induction heating at about 1150°C in a stainless steel (Inconel 601) crucible. The melt is finally cast into stainless steel canisters and lids are welded on. Because heat is evolved as a result of radioactive decay, the canisters must initially be air cooled to prevent the glass temperature from rising above 450°C. Eventually natural cooling is sufficient to prevent overheating.

All operations are, of course, carried out remotely and an essential feature of the plant design is the filtration and cleaning of the gas generated in the melter and calciner. The initial ten years of storage will be in a surface site, after which it is intended to move the containers to a safe underground store. In some proposals, the canisters will be surrounded by a clay buffer material to prevent contact between the canisters and any water which may enter the site.

Table 27 gives the glass composition, including the fission products and residues from fuel processing.

Table 27
Possible glass composition for solidification of LWR waste
(Laude 1985)

	weight %
SiO_2	45.2
B_2O_3	13.9
Al_2O_3	4.9
Na_2O	9.8
CaO	4.0
Fe_2O_3	2.9
NiO	0.4
P_2O_5	0.3
ZrO_2 (filings)	1.0
Li_2O	2.0
ZnO	2.5
Actinide oxides	0.9
Metallic particles	0.7

The composition is a compromise between a number of requirements:

1. Low corrosivity of the melting crucible and other furnace parts.
2. Viscosity at the casting temperature to be in the range 40–50 Pa s.
3. Low volatilisation during melting. Volatilisation of RuO_2 can be particularly troublesome but can be reduced by adding a reducing agent to the batch.
4. The glass must be effective as a containment medium for 10 000 years or more. Properties relevant to the long term stability of the glass are:
   a. No degradation in glass properties due to self-irradiation.
   b. Thermal stability. No devitrification at the maximum temperature likely to be encountered i.e. 450°C.
   c. Mechanical. Measures must be taken to prevent the glass fragmenting when the canisters are being moved or subjected to thermal stresses by forced cooling.
   d. Chemical stability. This has been assessed on blocks of active glass prepared under production conditions and by laboratory tests on powdered samples, using the 'distillation column' or Soxhlet method described in Chapter 2.

Laude quotes the following rates of extraction of active isotopes:

137_{Cs}	approx.	1.5×10^{-7} g.cm^{-2}.d^{-8}
90_{Sr}	approx.	5.0×10^{-8} g.cm^{-2}.d^{-8}
106_{Ru-Re}	approx.	5.0×10^{-8} g.cm^{-2}.d^{-8}
144_{Ce}	approx.	3.0×10^{-8} g.cm^{-2}.d^{-8}
Actinides		10^{-7} to 10^{-8} g.cm^{-2}.d^{-8}

Hench (1985a) points out that all the 27 nuclear waste glasses which have been studied worldwide fall within a narrow range of compositions. In Fig. 136 the oxides of Si, B and Na i.e. the sum of the major components of the non-radioactive frit is plotted on one side of the triangular composition diagram. The oxides of Fe, Al and all other constituents in the frit or in the waste (WP) are summed and taken as the third component. The numbered contours outline the areas of composition within which the leach rate is not more than the value given on the contour. All leaching experiments were carried out by the same method at 90°C and for a period of 28 days.

There is an appreciable area of compositions for which the leach rate is in the range 0.1 to 0.2 g m$^{-2} \cdot$d^{-1} but a very restricted region with leach rates as low as 0.02 g m$^{-2} \cdot$d^{-1}. All these glasses contain nearly equivalent 51–53 wt% SiO$_2$. Hench stresses the importance of the nature of the surface layer which is likely to form under typical storage conditions (Chapter 2). This should have a protective action against further leaching. The indications from controlled burial experiments is that this situation is likely to occur.

Wicks (1985) gives information on the quantities, compositions and physical characteristics of the various types of active waste currently stored in the USA. It is intended to follow the French practice of vitrification in a sodium borosilicate glass but with somewhat different compositions according to the origin of the waste. A considerable amount of information is given on chemical durability and the methods by which this has been studied. Many intercomparison tests were carried out using a standard technique in which the waste glass samples were inserted in various leachants contained in Teflon vessels. The solutions were analysed after various times, and experiments were carried out over a range of temperatures. Surface composition profiles were determined. Appropriate field tests were also carried out to study the leaching behaviour in underground storage conditions.

It is interesting (and fortunate) that the leachability decreases markedly with increasing waste product loading (Fig. 137).

It is suggested, following Hench, that the reduction in leaching rate is due to the formation of a protective layer rich in elements some of which are present in the waste e.g. Fe, Ni, Mn and some present in the glass frit, e.g. Mg, Ti, Zr.

The effect of pH on leach rate is similar to that well known for the simpler silicate glasses (Fig. 138).

The rate is low in the neutral range and fortunately the pH of most repository ground waters are likely to be in this range. The pH will tend to rise as the ground water attacks the glass but it is suggested that the rise will not be large because of buffering effects of package components and of constituents of the ground water.

The effect of temperature has been studied up to 350°C. At the highest temperatures the attack is considerable but under the expected repository conditions, it is not expected that the glass-water interface temperature will rise above 100°C. Under such conditions the glass durability is considered to be sufficiently good.

In the early stages of leaching, the quantity of material extracted increases as t$^{1/2}$ but at long times the rate becomes constant. Thus if the surface reaction layer remains undisturbed, a short term laboratory test is likely to overestimate the total long term extraction.

Wicks reviews the available information on fabrication stresses, impact resistance, thermal stability and radiation effects, concluding that there are no grounds for concern on these accounts.

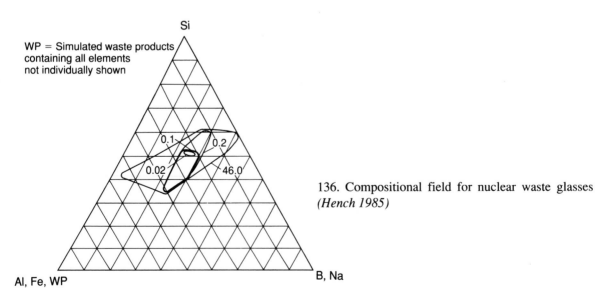

136. Compositional field for nuclear waste glasses *(Hench 1985)*

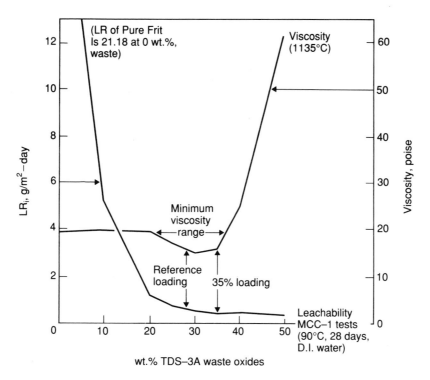

137. Leachability and viscosity of waste glass as a
function of waste loading
(Wicks 1983)

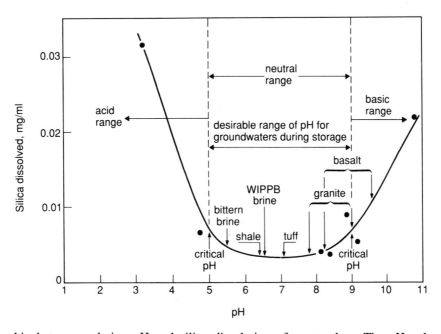

138. Relationship between solution pH and silica dissolution of waste glass. The pH values are for the
groundwaters of geological formations considered suitable as waste disposal sites
(Wicks 1985)

His review concludes with U.S. Department of Energy estimates of the radiation exposures associated with this method of storage (Table 28). Some arise in the vitrification process and some during transport to the storage site. The long term storage hazards are estimated on the assumption that the ground water will be prevented from coming into contact with the glass for hundreds and possibly thousands of years. Even then, it is suggested that it would take a similar period of time before any active species could reach the accessible environment.

Table 28
Projected defence waste processing facility risks: Transportation, geological burial

Waste recovery and immobilisation		
(Max. industrial dose)		
Normal operation	0.2	mrem/yr.
Postulated accidents	0.04	mrem/yr.
Transportation		
(Radiologic –max. individual dose)		
Normal (per shipment)	0.0006	mrem/yr.
Accident: rail	7.5	mrem/yr.
Accident: truck	1.5	mrem/yr.
Geological disposal (peak dose)		
1.6 km well (Max. individual)	0.06	mrem/yr.
River system (avg. Individual)	0.000003	mrem/yr.

Normal background radiation levels		
Savannah River Plant, Aika, South Carolina	93	mrem/yr.
Atlanta, Georgia	124	mrem/yr.
Denver, Colorado	195	mrem/yr.

Other sources of radiation		
One chest X-ray	20	mrem
Brick house	5	mrem/yr.
Television		0.2 to 1.5 mrem/yr.

He comments at one point in his review that, although storage in glass is likely to be used in many countries, storage in a crystalline material may be preferred in future. This point is discussed in detail in Hayward's review (1988). He considers in detail the respective merits of a number of chemically durable glass–ceramics for waste storage. The materials discussed include compositions related to the minerals celsian, fresnoite, basalt and sphene. Substantial research programmes are in progress on these materials in a number of countries. As might be expected, they have superior thermal and mechanical properties to the borosilicate glasses and some compositions have leaching rates which are 10 to 1000 times smaller. A factor which may have weighed against their earlier acceptance is the higher melting temperatures needed.

14.4. Bioactive glasses and glass-ceramics

A feature of modern surgery is the widespread use of many different materials as implants. Hench and his colleagues at the University of Florida have been very active in the study of various aspects of the use of glass in this field: the development of suitable glass compositions, the study of the mechanisms responsible for forming a bond between the glass surface and either soft body tissue or bone and in carrying out programmes of *in vitro* and *in vivo* trials. The latter have included the successful insertion of implants in teeth and in the middle ear (Wilson 1985) and in bone (Hench and Spilman 1985b). Depending on the nature of the application, the glass may be used as such or as a coating on metal or ceramic. The materials are marketed under the name Bioglass™.

Hench's materials are essentially Na_2O–CaO–SiO_2 glasses with a low SiO_2 content and containing about 6-wt% of P_2O_5. The most effective glasses lie within area A in Fig. 139. A typical composition is 45%SiO_2, 24.5%CaO, 24.5%Na_2O, 6%P_2O_5 (Pantano *et al.* 1974).

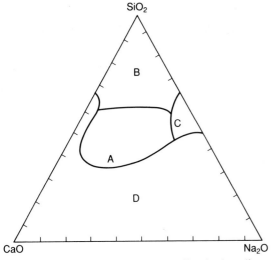

139. 'Bioglass' bone-bonding composition diagram (with constant wt% P_2O_5)
Region (A). Bonding within 30 days
(Pantano et al. 1974)

Hench and Spilman (1985) describe the bonding mechanism as follows:

Upon implantation of Bioglass™ devices, there is a rapid exchange of alkali ions from the glass surface with H^+ or H_3O^+ ions from body fluids. This exchange results in the formation of a highly biologically active silica–rich layer on the Bioglass™ surface while at the same time maintaining an alkaline pH at the implant interface. The consequence of high bioreactivity and alkaline pH is the development of a CaO–P_2O_5 rich layer within minutes to hours of implantation. Nucleation of a CaO–P_2O_5 layer occurs on the Bioglass™ surface but growth proceeds by incorporation of Ca and P ions from body fluids as well as from the glass network. Consequently an anionic balance of $-OH^-$ and $-CO_3^-$ groups characteristic of natural tissues is achieved within the growing CaO–P_2O_5 layer. Within a 1 to 2 week period of time, crystallisation of the calcium phosphate layer into a hydroxy–carbonate–apatite proceeds to completion.

Interesting results have also been obtained by Vogel and his collaborators using a machinable bioactive glass ceramic (Höland *et al.* 1983, 1985). There obviously can be advantages in being able to machine the implant to shape just before insertion. Also the potentially higher strength or toughness of a glass–ceramic can be an important advantage.

The materials were made from a glass with a composition in the range: 19–52% SiO_2, 12–23% Al_2O_3, 5–15% MgO, 9–30% CaO, 3–10% K_2O/Na_2O, 0.5–7 F^- and 4–24% P_2O_5. Heat treatment is carried out in the temperature range 610°–1050°C and results in the formation of two crystalline phases–apatite and mica, with some residual glassy phase.

Bonding of the implant to bone involves the formation of a calcium phosphate layer 5–10 μ thick between the bone and the implant after a reaction time of 16 weeks. The bending strength of the glass–ceramic is about 200 MPa and the shear strength between bone and biomaterial is about 4 MPa.

14.5. Controlled release glasses

The compositions and application of controlled release glasses are described in a number of publications and patents (Roberts 1975; Drake 1978, 1983; Telfer *et al.* 1986; Knott 1989). The more important applications are to remedy trace element deficiencies in plants and animals. The elements to be released include iron, cobalt, copper and selenium and the solvent glass is usually a phosphate composition. The rate at which the appropriate element is released depends on the glass composition and the nature of the aqueous medium to which it is exposed. Lengthy field trials are needed to ensure that the release rate is not excessive but is sufficient to be effective over the required period of time.

Roberts (1975) describes the application of glasses in the FeO–K_2O–P_2O_5 system to supply micronutrient quantities of ferrous iron. Field tests showed that application of the glass frit to alkaline soils at the rate of 25lb per acre greatly increased the yield of tomatoes and at 8lb per acre iron chlorosis was eliminated from corn crops.

Knott and his colleagues have been primarily concerned with the treatment of animals for trace element deficiencies. The glass is either implanted subcutaneously or fed as a bolus. Extensive and careful animal trials were carried out. For animals grazing on poor pastures, significant increases in liveweight were recorded for the treated compared with the untreated animals. This method of treatment is particularly attractive for animals which range over wide areas, in that it eliminates or greatly reduces the need to herd the animals, as is required by other methods of treatment, e.g. injection.

References

Abendroth, R.P.: Oxide formation and adherence on an iron-cobalt-nickel glass sealing alloy. Mat. Res. Stand. **5** (1965) 459–466.

Abou-el-Leil, M.and Cooper, A.R.: Analysis of field-assisted binary ion exchange. J. Am. Ceram. Soc. **62** (1979) 390–395.

Adam, H.: The theoretical foundations of glass compression seals and their practical consequences. J. Soc. Glass Technol. **38** (1954) 285–296.

Adler, D., Shur, M.S.; Silver, M.and Ovshinsky, S.R.: Threshold switching in chalcogenide glass films. J. Appl. Phys. **51** (1980) 3289–3309.

Andrews, E.H., He Pingsheng. and Vlachos, C.: Adhesion of epoxy resin to glass. Proc. R. Soc. **A381** (1982) 345–360.

Anon: A standard method of determining stresses in glass–to–metal seals of the sandwich and bead types. J. Soc. Glass Technol. **53** (1949) 77–87.

Anon: 500 bottles per minute by the ribbon machine. Glass Ind. **49** (1968) 550–5.

Anon: Handling hot containers with a polimide composite material. Glass Technol. **20** (1979) 49–50.

Anon: Joint venture formed to develop container coating. Glass Ind. **69** (12) (1988a) 10–14.

Anon: Keynote report on packaging (Glass). Keynote Publications, London (1988b).

Araujo, R.J.: Photochromic glass. In: Tomozawa, M. and Doremus, R.H. (eds) 'Treatise on materials science and technology. Vol 12: Glass I. Interactions with electromagnetic radiations'. Academic Press. London, New York (1977) 91–122.

Araujo, R.J.: Photochromism in glasses containing silver halides. Contemp. Phys. **21** (1980) 77–84.

Araujo, R.J.: Photosensitive glasses In: Boyd, D.C. and MacDowell J.F.(eds): 'Commercial Glasses' American Ceramic Society, Columbus OH (1986) 151–156.

Arfsten, N.J.: Sol–gel derived transparent IR-reflecting ITO semiconductor coatings and future applications. J. Non-Cryst. Solids **63** (1984) 243–250.

Armistead, W.H. and Stookey, S.D.: Photochromic glasses sensitized by silver halides. Science **144** (1964) 150–158.

Ashbee, K.H.G.: Fundamental properties of fiber reinforced composites. Technomic Publishing Co. Lancaster PA (1989) 372 pp.

ASTM: Designation F144–73. Reference glass–metal sandwich seal and testing for expansion characteristics by polarimetric methods. ASTM Book of Standards. Part 43 (1977a).

ASTM: Designation F14–68. Reference glass–metal bead seal. ASTM Book of Standards. Part 43 (1977b).

ASTM: Designation F140–73. Reference glass–metal butt seal. ASTM Book of Standards. Part 43 (1977c).

ASTM C147–76 Standard method of internal pressure test on glass containers.

ASTM C149–77 Standard method of thermal shock test on glass containers.

Aubourg, P.F. and Wolf, W.W.: Glass Fibers. In: Boyd, D.C. and MacDowell J.F.(eds): 'Commercial Glasses' American Ceramic Society, Columbus OH (1986) 51–64.

Augustsson, B.O., Wasylyk, J.S., Smay, G.L. *et al.*: Computer modelled internal pressure strength predictions for refillable glass containers. Glastech. Ber. **59** (1986) 121–131.

Badger, A.E., Weyl, W. and Rudow, H.: Effect of heat treatment on the colour of gold ruby glass. Glass Ind. **20** (1939) 407–414.

Bamford, C.R.: A theoretical analysis of the optical properties of 'Spectrafloat' glass. Physics Chem. Glasses **17** (1976) 209–213.

Bamford, C.R.: Application of the ligand field theory to coloured glasses. Physics Chem. Glasses **3** (1962) 189–202.

Bamford C.R.: Colour generation and control in glass. Elsevier. Amsterdam, Oxford, New York (1977).

Bansal, N.P. and Doremus, R.H: Handbook of Glass Properties. Academic Press. New York, London (1986) 680 pp.

Barnett, M.: Temperature distributions in double glazing with aluminium and timber frames. BSE **44** (1977) 250–3.

Barsom, J.M.: Fracture of tempered glass. J. Am. Ceram. Soc. **51** (1968) 75–78.

Bartholomew, R.F. and Garfinkel, H.M.: Chemical strengthening of glass. In: Uhlmann, D.R. and Kreidl, N.J. (eds) 'Glass: Science and Technology. Vol 5. Elasticity and Strength in Glass'. Academic Press. New York (1980) 217–270.

Bates, T: Ligand field theory and absorption spectra of transition metal ions in glass. In: Mackenzie, J.D. (ed.) 'Modern Aspects of the Vitreous State Vol. 2'. Butterworth. London (1962) 195–224.

Baucke, F.G.K.: Development of an electrochromic mirror. In: Wright, A.F. and Dupuy, J. (eds). 'Glass... Current issues. NATO ASI series. Series E: Applied Sciences – No. 92'. Martinus Nijhoff. Dordrecht, Boston, Lancaster (1985) 506–518.

Baucke, F.G.K. and Werner, R.D.: Mixed alkali effect of electrical conductivity in glass-forming silicate melts. Glastech. Ber. **62** (1989) 182– 86.

Baynton, P.L., Rawson, H. and Stanworth, J.E.: The semiconducting properties of some vanadate glasses. J. electrochem. Soc. **104** (1957) 237–240.

Beales, K.J. and Day, C.R.: A review of glass fibres for optical communications. Physics Chem. Glasses **21** (1980) 5–21.

Beall, G.H.: Structure, properties and applications of glass–ceramics. In: 'Advances in Nucleation and Crystallization in Glasses'. Am.Ceram.Soc. Special Publication No. 5 (1971) 251–261.

Beall, G.H.: Microstructure of glass ceramics and photosensitive glasses. Glass Technol. **19** (1978) 109–113.

Beall, G.H. and Duke, D.A.: Glass–ceramic technology. In: Uhlmann, D.R. and Kreidl, N.J. (eds). 'Glass. Science and Technology Vol. 1. Glass-forming systems'. Academic Press. New York, London, San Francisco (1983) 404-446.

Beall, G.H.: Property and process development in glass-ceramic materials. In: Wright, A.F. and Dupuy, J. (eds) 'Glass... Current issues. NATO ASI series. Series E: Applied Sciences – No. 92'. Martinus Nijhoff. Dordrecht, Boston, Lancaster (1985) 31–48.

Beall, G.H.: Glass–ceramics. In: Boyd, D.C. and MacDowell J.F.(eds) 'Commercial Glasses' American Ceramic Society, Columbus OH (1986) 157–173.

Bean, K.E.: CVD applications in micro-electronic processing. Thin Solid Films **83** (1981) 173–186.

Beason, W.L. and Morgan, J.E.: Glass failure prediction model. J. Struct. Engrg. ASCE **110** (1984) 197–212.

Beason, W.L.: Structural analysis of sealed insulating glass. J. Struct. Engrg. ASCE **112** (1986) 1133–1146.

Behr, R.A., Minor, J.E. and Linden, M.P.: 'Load duration and interlayer thickness in laminated glass' J. Struct. Engng. **112** (1986) 1441–1453.

Beier, W. and Frischat, G.H.: Transport mechanisms in alkali silicate glasses. J. Non-Cryst. Solids **73** (1985) 113–133.

Berg, C.A.: The motion of cracks in plane viscous deformation. In: Proc.4th National Congress on Applied Mechanics. Vol.2 (1962) 885–892.

Blackman, L.C.F.: Development of glass fibre reinforced cement. Proc. Xth. Internat. Congr. on Glass, Kyoto. (1974) Section 10. 69–76.

Blackman, L.C.F.: Glass fibre-reinforced cement: a progress report. Composites **10** (1979) 69–72.

Blandenet, G. , Court, M. and Lagarde, Y.: Thin layers deposited by the Pyrosol process. Thin Solid Films **77** (1981) 81–90.

Blizard, J.R. and Howitt, J.S.: Development of a windshield. Glass Industry **50** (1969a) no.12. 573–5; **51** (1970b) no.1. 16–18; (1970c) no.2. 73–75.

Blocher Jr, J.M.: Coating of glass by CVD. Thin Solid Films **77** (1981) 51–64.

Bockris, J. O'M. and Lowe, D.C.: Viscosity and structure of molten silicates. Proc. R. Soc. A 226 (1954) 423–435.

Bockris, J. O'M, Mackenzie, J.D. and Kitchener, J.A.: Viscous flow in silica and binary liquid silicates. Trans. Faraday Soc. **51** (1955) 1734.

Bornstein, A., Croitoru, N. and Marom, E.: Chalcogenide glass fibres. J.Non-Cryst. Solids **74** (1985) 57–65.

Borrelli, N.F., Chodak, J.B, Nolan, D.A. and Seward, T.P.: Interpretation of induced colour in polychromatic glasses. J. Opt. Soc. Amer. **69** (1979) 1514–1519.

Borrelli, N.F., Morse, D.L., Bellman, R.H. *et al.*: Photolytic technique for producing microlenses in photosensitive glass. Applied Opt. 24 (1985) 2520–2525.

Bourne, R.; Cowan, N.D. and Budd, S.M.: Damage to glass surfaces at elevated temperatures: Effects of hardness and thermal conductivity of damaging material. Glass Technol. 25 (1984) 145–147.

Boyd, D.C. and MacDowell J.F.(eds): 'Commercial Glasses' American Ceramic Society, Columbus OH (1986) 229 pp.

Bray, P.J. and O'Keefe, J.G.: Nuclear magnetic resonance investigations of the structure of alkali borate glasses. Physics Chem. Glasses 4 (1963) 37–46.

Bretschneider, J.: Application of optical testing procedures to quality control of flat glass. Glastech. Ber. **61** (1988) 172–175.

Brinker, C.J., Clark, D.E. and Ulrich, D. (eds): 'Better Ceramics through Chemistry' North Holland (1984a).

Brinker, C.J., Drotning, W.D. and Scherer, G.W.: A comparison between the densification kinetics of colloidal and polymeric silica gels. In: Brinker, C. J., Clark, D. E. and Ulrich, D. R. (eds) 'Better Ceramics through Chemistry' North Holland (1984b) 25–32.

British Standards Institution: CP3 Code of basic data for the design of buildings. (1972) Chapter V. Part 2. Wind loads.

British Standards Institution: BS 5051. Part 1 (1973). Bullet resistant glazing for interior use.

British Standards Institution: BS 5051. Part 2 (1979). Bullet proof glazing for exterior use.

British Standards Institution: BS 1376. (1974). Specification for colours of light signals.

British Standards Institution: BS 5466. Part 1 (1977). Methods for corrosion testing of metallic coatings. Neutral spray test. (Equivalent to ISO 3768: 1976.).

British Standards Institution: BS 5466. Part 3 (1977). Methods for corrosion testing of metallic coatings. Copper-accelerated acetic acid salt spray test. (Equivalent to ISO 3770: 1976.).

British Standards Institution: BS 5713. (1979). Specification for hermetically sealed flat double glazing units.

British Standards Institution: BS AU 178. (1980). Road vehicle safety glass.

British Standards Institution: BS 6206. (1981). Impact performance requirements for flat safety glass and safety plastics for use in buildings.

British Standards Institution: BS 6118. (1981). Multitrip glass bottles for beer and cider.

British Standards Institution: BS 6119. Part 1 (1981). Specification for multitrip 750 ml and 1 litre bottles.

British Standards Institution: BS 6262. (1982). British standard code of practice for glazing in buildings.

British Standards Institution: BS 1679. Part 6 (1984). Containers for pharmaceutical dispensing.

British Standards Institution: BS 5357. (1985). Code of practice for the installation of security glazing.

British Standards Institution: BS 7034 (or ISO 7884: 1987). Viscosity and viscometric fixed points of glass.

British Standards Institution: BS 874. Methods for determining thermal insulating properties. Part 2. Section 2.1. (1987) Guarded hot box method.

British Standards Institution: BS 6375. Part 1 (1989). Performance of windows.

British Standards Institution: BS 6993. Part1 (1989). Thermal and radiometric properties of glazing. Calculation of U values.

British Standards Institution: BS 6993. Part 2 (1990). Thermal and radiometric properties of glazing. Direct measurement of glazing U values.

Brungs, M.P.: Assessment of polyurethane coatings as a means of maintaining the strength of lightweight containers. Glass Technol. **29** (1988) 103–5.

Budd, S.M. and Cornelius, W P.: Impact studies on glass containers. Glass Technol. **17** (1976) 170–3.

Budd, S.M.; Cowan, N.D. and Bourne, R.: Damage to glass surfaces by various materials at elevated temperatures. J. Non-Cryst. Solids **38/39** (1980) 409–412.

Budd, S.M.: Real world glass strength- as disclosed by tests on glass containers. In: Kurkjian, C.J. (ed.) 'Strength of Inorganic Glass. NATO Conference Series. Series VI. Materials Science. Volume 11'. Plenum Press. New York (1985a) 419–422.

Budd, S.M.: Evaluation of carbon-fibre reinforced composites for handling glass containers. Glass Technol. **26** (1985b) 245–247.

Budd, S.M.: Functional coatings for glass containers. Glass Technol. **28** (1988) 227–31.

Burggraaf, A.J. and Cornellisen, J.: The strengthening of glass by ion exchange. Part 1. Stress formation by ion diffusion in alkali aluminosilicate glasses. Physics Chem. Glasses **5** (1964) 123–129.

Burggraaf, A.J.: The strengthening of glass by ion exchange. Part 2. Stress formation and stress relaxation after ion exchange in alkali aluminosilicate glasses in connection with structural changes in the glass. Physics Chem. Glasses **7** (1966) 169–172.

Cahn, R.W.: Metallic glasses. Contemp. Phys. **21** (1980) 43–75.

Cayless, M.A. and Marsden, A.M. (eds): Lamps and Lighting. 3rd Edn. (1983) Edward Arnold. London, Baltimore.

Charles, R.J.: The mixed alkali effect in glasses. J. Am. Ceram. Soc. **48** (1965) 432–3.

Charles, R.J.: Metastable liquid immiscibility in alkali metal oxide- silica glasses. J. Am. Ceram. Soc. **49** (1966) 55–62.

Charles, R.J.: Static fatigue of glass I. J. Appl. Phys. **29** (1958a) 1549–1553.

Charles, R.J.: Static fatigue of glass II. J. Appl. Phys. **29** (1958b) 1554–1560.

Charles, R.J.: Dynamic fatigue of glass. J. Appl. Phys. **29** (1958c) 1657–1662.

Charles, R.J.: A review of glass strength. In: Burke, J.E. (ed.) 'Progress in ceramic science. Vol. 1'. Pergamon Press. Oxford, London, New York. (1961) 1–38.

Charnock, H.: The float glass process. Phys. Bull. **21** (1970) 151–156.

Chen, A.C.M., Dunham, A.M. and Wang, J.M.: Electron beam readout sensitivity of Ge–Te–X amorphous thin films. Proc. 5th Int. Congr. Amorphous and Liquid Semiconductors. Taylor and Francis, London (1973) 701–706.

Chludzinski, P., Ramaswamy, R.V. and Anderson, T.J.: Ion exchange between soda–lime–silica glass and molten sodium nitrate-silver nitrate molten salts. Physics Chem. Glasses **28** (1987) 169–173.

Chopra, K.L. and Pandya, D.K.: Transparent conductors. A status review. Thin Solid Films. **102** (1983) 1–46

Christopher, E.C. and Murialdo, G.: A novel forced convection cooling system. Glass **54** (1977) 305–311.

Chyung, C.K., Beall, G.H. and Grossmann, G.: Fluorphlogopite mica glass ceramics. Proc. Xth Internat. Congr. on Glass Kyoto. (1974) Section 14. 33–40.

CIBSE Guide A1. Section on thermal comfort.

Clark, D.E., Pantano, C.G. and Hench, L.L.: 'Corrosion of glass'. Magazines for Industry Inc. New York (1979).

Clarkin, J.P., Skutnik, B.J. and Munsey, B.D.: Enhanced strength and fatigue resistance of silica fibres with hard polymeric coatings. J.Non-Cryst. Solids **102** (1988) 106–111.

Clasen, R.: Preparation of high purity glasses by sintering of colloidal particles. Glastech. Ber. **60** (1987a) 125–132.

Clasen, R.: Preparation and sintering of high density green bodies to high–purity silica glasses. J. Non-Cryst Solids **89** (1987b) 335–344.

Cohen, M.H. and Turnbull, D.: Composition requirements for glass formation in metallic and ionic systems. Nature (London) **189** (1961) 131.

Coney, S.S.: The selection and use of coatings for glass forming equipment. Glass Technol. **19** (1978) 144–8.

Cook, R.F.: Economics of glass recycling. Glass **63** (1986) 'Glassman 86' C53–54.

Cornelissen, J. and Zijlstra, A.L.: The strength of glass rods as a result of various treatments. In: 'Symposium sur la resistance mecanique du verre et les moyens de l'ameliorer'. USCV, Charleroi (1961) 337–358.

Dale, A.E. and Stanworth, J.E.: A note on some very soft glasses and some of their applications. J. Soc. Glass Technol. **32** (1948) 147–153.

Dale, A.E. and Stanworth, J.E.: The development of some very soft glasses.J. Soc. Glass Technol. **33** (1949) 167–175.

Davis, E.A. and Shaw, R.F.: Characteristic phenomena in amorphous semiconductors. J. Non-Cryst. Solids **2** (1970) 406–431.

Davies, H.A.: The formation of metallic glasses. Physics Chem. Glasses **17** (1976) 159–173.

Davies, M.W., Kerrison, B., Gross, W.E. *et al.*: Slagceram: a glass ceramic from blast furnace slag. J. Iron and Steel Inst. **208** (4) (1970) 348–370.

Davis, L.A.: Mechanical responses of metallic glasses. In: Wright, A.F. and Dupuy, J. (eds) 'Glass...current issues. NATO ASI Series. Series E: Applied Sciences – No. 92'. Martinus Nijhoff Publishers. Dordrecht, Boston, Lancaster (1985) 94–124.

Davy, J.R.: Development of calcium aluminate glasses for use in the infrared to 5 μm. Glass Technol. **19** (1978) 32–36.

De Bie, J. and McBride, M.W.: The evolution of DC sputter coating. Glass Production Technology International (1991) 185–188.

Deeg E. W. Optical glasses. In: Boyd, D.C. and MacDowell J.F.(eds): 'Commercial Glasses' American Ceramic Society, Columbus OH (1986) 9–34.

Department of Energy (UK): Energy Efficiency Booklet 7. Degree Days.

Dislich, H.: New routes to multicomponent oxide glasses. Angew. Chem. Internat. Edn. **10** (1971) 363–370.

Dislich, H. and Hussmann, E.: Amorphous and crystalline dip coatings obtained from organometallic solutions. Procedures, chemical processes and products. Thin solid films **77** (1981) 129–139.

Dislich, H.: Glassy and crystalline systems from gels: chemical basis and technical application J. Non-Cryst. Solids **63** (1984) 237–242.

Dislich, H., Hinz, P., Arfsten, N-J. and Hussmann, E.: Sol–gel yesterday, today and tomorrow. Glastech. Ber. **62** (1989) 46–51.

Doremus, R.H.: Optical properties of small gold particles. J. Chem. Phys. **40** (1964) 2389–96.

Doremus, R.H.: Optical properties of small silver particles. J. Chem. Phys. **42** (1965) 414–7.

Doremus, R.H.: Chapter 1 in: Marinsky, J.A. (ed.) 'Ion exchange – a series of advances, Vol.2'. Marcel Dekker. New York (1969).

Doremus, R.H.: Glass Science. Wiley. New York (1973).

Dorn, R.; Baumgartner, A. *et al.*: Glass fibres from mechanically shaped preforms. Glastech. Ber. **60** (1987) 79–82.

Douglas, R.W. and El Shamy, T.M.M.: Reaction of glasses with aqueous solutions. J. Am. Ceram. Soc. **50** (1967) 1–7.

Douglas, R.W. and Frank, S.: A history of glass. G.J.Forbes. Henley-on-Thames (1972).

Doyle, P.J.(ed.): Glassmaking today. Portcullis Press. Redhill (1979).

Dow-Corning: British Patent 1 454 960 (1974).

Downs, R.L.: Hollow glass microspheres by sol–gel technology. In: Klein, L.C. (ed.) 'Sol–gel technology for thin films, fibers, preforms and speciality shapes' Noyes Publications (1988) 330–381.

Drake, C.F. and Tripp, M.: UK Patent GB 2 037 735. (1983).

Drake, C.F.: UK Patent 1 512 637. (1978).

Drexhage, M.G.: Heavy metal fluoride glasses. In: Tomozawa, M.; Doremus, R. (eds) 'Treatise on Materials Science and Technology. Volume 26, Glass IV'. Academic Press. New York, London (1985) 151–243.

Dumbaugh, W.H.; Flannery, J.E. and Megles, J.E.: Strong composite glasses. J. Non-Cryst. Solids **38/38** (1980) 469–474.

Duwez, P., Willens, R.H. and Klement, W.: Continuous series of metastable solid solutions in silver–copper alloys. J. Appl. Phys. **31** (1960) 1136.

Duwez, P. and Lin, S.C.H.: Amorphous ferromagnetic phase in iron–carbon–phosphorus alloy. J. Appl. Phys. **38**(1967) 4096–4097.

Edge, C.K.: Flat Glass Manufacturing Processes. In: Tooley, F.V. (ed.) 'Handbook of Glass Manufacture. Vol II. 3rd Edn'. Ashlee Publishing Co Inc. New York (1984).

Edge, C.K.: Float Glass. In: Boyd, D.C. and MacDowell J.F.(eds): 'Commercial Glasses' American Ceramic Society, Columbus OH (1986) 43–50.

Elliott, S.R.: 'Physics of amorphous materials'. Longman. London, New York (1983).

Elliott, S.R.: The use of non-diffraction probes in determining the structure of amorphous solids. In: Proc. XVth Internat. Congr. on Glass, Leningrad (1989) 65–83.

Elyard, C.A. and Rawson, H.: The resistance of glasses of simple composition to attack by sodium vapour at elevated temperatures. In: 'Advances in Glass Technology. Vol. I' Plenum Press. New York (1962) 270–286.

Ensor, T.: Mould materials. Glass Technol. **19** (1978) 113–119.

Ernsberger, F.M.: Detection of strength impairing flaws in glass. Proc. R. Soc. **A257** (1960) 213–223.

Ernsberger, F.M.: A study of the origin and frequency of occurrence of Griffith microcracks on glass surfaces. In: 'Advances in Glass Technology. Vol. I' Plenum Press. New York (1962) 511–524.

Ernsberger, F.M.: Strength of brittle ceramic materials. Am. Ceram. Soc. Bull. **52** (1973) 240–246.

Espe, W.: 'Materials of high vacuum technology'. Pergamon. Oxford, New York (1968).

Estrada, W., Andersson, A.M. and Granqvist, C.G.: Electrochromic nickel oxide based coatings made by dc magnetron sputtering: Preparation and optical properties. J. Appl. Phys. **64** (1988) 3678–3683.

European Pharmacopoeia (1971): Procedure for the determination of the hydrolytic resistance of the inside of glass containers.

'European solar radiation atlas. Vol 2. Inclined surfaces'. Ed. W.Polz. Verlag TÜV Rheinland. EUR 9344. (1984).

Eversteijn, F.C.; Severin, P.J.W., Van der Brekel, C.H.J. and Peek, H.L.: A stagnant layer model for the epitaxial growth of silicon from silane in a horizontal reactor. J. Electrochem. Soc. **117** (1970) 925–931.

Fainaro, I., Ish Shalom, M. and Ron, M.; Lipson, S.: Interdiffusion of silver in glasses and the related variations in electronic polarisability. Physics Chem. Glasses **25** (1984) 16–21.

Fairbanks, H.V.: Effects of surface conditions and chemical composition of metal and alloys on the adherence of glass to metal. In: 'Symposium sur le contact du verre chaud avec le métal'. U.S.C.V., Charleroi (1964) 573–595.

Fan, L.T. and Fairbanks, H.V.: Radioactive tracer studies on the mechanism of adherence of glass to metal.. In: 'Symposium sur le contact du verre chaud avec le métal'. U.S.C.V., Charleroi (1964) 621–637.

Fan, J.C.C., Bachner, F.J., Foley, G.H. and Zavracky, P.M.: Transparent heat–mirror films of $TiO_2/Ag/TiO_2$ for solar collection and radiation insulation. Appl. Phys. Lett. **25** (1974) 693–695.

Fellows, C.J. and Shaw,F.: A laboratory investigation of glass to mould heat transfer during pressing. Glass Technol. **19** (1978) 4–9.

Findakly, T.: Glass waveguides by ion exchange – a review. Opt. Eng. **24** (1985) 244–250.

Fleming, G.R.: Ovonic electroluminescent arrays. J. Non-Cryst. Solids **2** (1970) 540–549.

Forbes, D.W.A.: Solder glass seals in semiconductor packaging. Glass Technol.**8** (1967) 32–42.

France, P.W., Carter, S.F., Moore, M.W., Williams, J.R. and Day, C.R.: Status and prospects for mid-infrared fibres. In: 'ECOC '88. 14th European Conf. on Optical Communications. (Conf. Pub. No. 292). Vol.1'. Brighton, Sept. 1988. 428–432.

Francel, J.: Sealing glasses. In: Boyd, D.C. and MacDowell J.F.(eds) 'Commercial Glasses' American Ceramic Society, Columbus OH (1986) 79–86.

Frank, G., Kauer, E. and Köstlin, H.: Transparent heat-reflecting coatings based on highly doped semiconductors. Thin Solid Films **77** (1981) 107–118.

Frank, W.R.B.: A review of recent work on the working temperature of glassware moulds in various countries. Glass Technol. **2** (1961) 107–110.

Fréchette, V.D.: The fractology of glass. In: Stevens, H.J. and LaCourse, W.C. (eds) 'Introduction to glass science'. Plenum Press. New York, London (1972) 433–450.

Fréchette, V.D. and Michalske, T.A.: Fragmentation in the bursting of containers. Am. Ceram. Soc. Bull. **57** (1978) 427–429.

Freiman, S.W.: Fracture mechanics of glass. In: Uhlmann ,D.R. and Kreidl, N.J. (eds). 'Glass: Science and Technology. Vol. 5. Elasticity and Strength of Glass'. Academic Press. New York, London. (1980) 21–78.

French, W.G. and Pearson, A.D.: Refractive index changes produced in glass by ion exchange. Am. Ceram. Soc. Bull. **49** (1970) 974–977.

Fremaux, J. and Sauvinet, V.: Dépôt en continu de couche à bas émissivité sur verre flotté par pyrolyse de poudre. Rev. della Staz. Sper. Vetro **16** (1986) 693–697.

Frick, W.: Chemische Abscheidung von Nickel–Borschichten und ihre Anwendung in der Hohlglasindustrie. Glastech. Ber. **55** (1982). 13–18.

Fricke, J.: Thermal insulation materials from the sol–gel process. In: Klein, L.C. (ed.) 'Sol–gel technology for thin films, fibers, preforms and speciality shapes'. Noyes Publications (1988) 226–246.

Frieser, R.G.: A review of solder glasses. Electrocomm. Sci. Technol. **2** (1975) 162–199.

Fujiura, K.; Ohishi, Y. and Takahashi, S.: Organometallic chemical vapour deposition of ZrF_4-based fluoride glasses. Jap. J. Appl. Phys. **28** (1989) L147–149.

Gabano, J.P.: Applications of glasses in all solid state batteries. In: Wright, A.F. and Dupuy, J. (eds) 'Glass... Current issues. NATO ASI series. Series E: Applied Sciences – No. 92'. Martinus Nijhoff. Dordrecht, Boston, Lancaster (1985) 457–480.

Gambino, R.J.: Applications of thin film amorphous magnetic materials. In: Wright, A.F. and Dupuy, J. (eds) 'Glass... Current issues. NATO ASI series. Series E: Applied Sciences – No. 92'. Martinus Nijhoff. Dordrecht, Boston, Lancaster (1985) 153–171.

Gambke, T. and Metz, B.: Electrochromic layers for active optical filters. Glastech. Ber. **62** (1989) 38–45.

Gambling, W.A.: The evolution of low loss optical fibres. Physics Chem. Glasses **21** (1980) 1–4.

Gambling, W.A.: Glass, light and the information revolution. Glass Technol. **27** (1986) 179–187.

Gardon, R: Thermal tempering of glass. In: Uhlmann, D.R. and Kreidl, N.J. (eds) 'Glass Science and Technology. Vol. 5. Elasticity and Strength of Glass'. Academic Press. New York, London (1980) 146–217.

Gardon, R.: Evolution of theories of annealing and tempering: historical perspective. Am. Ceram. Soc. Bull. **66** (1987) 1594–9.

Garfinkel, H.M.: Ion-exchange equilibria between glass and molten salts. J. Phys. Chem. **72** (1968) 4175–4181.

Garfinkel, H.M.: Strengthening of glass by ion exchange. Glass Industry (1969) 28–31, 74–76.

Garfinkel, H.M. and King, C.B.: Ion concentration and stress in a chemically tempered glass. J. Am. Ceram. Soc. **53** (1970) 686–691.

Garfinkel, H.M.: Some recent new applications of glass technology. J. Non-Cryst. Solids **80** (1986) 69–82.

Geffcken, W.: Dünne Schichten auf Glas. Glastech. Ber. **24** (1951) 143–151.

Gehloff, G. and Thomas, M.: Die physicalischen Eigenschaften der Gläser in Abhängigkeit von der Zusammensetzung. Z. tech. Physik **6** (1925) 544–554 and **7** (1926) 105–126, 260–278.

Geittner, P. and Lydtin, H.: Manufacturing optical fibres by the PCVD process. Philips Tech. Rev. **44** (1989) 241–249.

Geotti-Bianchini, F. and Polato, P.: Solar control and thermal insulation with coated float glass. Proc. 2nd Int. Congress for Surface Technology. (1983). VDE Verlag Berlin-Offenbach. 361–369.

Giegerich, W. and Trier, W.: 'Glass machines'. Springer, Berlin. (1969).

Gillery, F.H.: Optical properties of coated flat glass. J. Non-Cryst. Solids. **47** (1982) 21–26.

Gläser, H.J.: Verfahren zur Beschichtung von Fensterscheiben mit Sonnen- und Wärmeschutzschichten. Glastech.Ber. **53** (1980a) 245–258.

Gläser, H.J.: Improved insulating glass with low emissivity coatings based on gold, silver or copper embedded in interference layers. Glass Technol. **21** (1980b) 254–261.

Gläser, H.J.: Beschichtung grossflächiger Glasscheiben durch Kathodenzerstäubung. Glastech. Ber. **56** (1983) 231–240.

Gläser, H.J.: Coated insulating glass. Glastech. Ber. **62** (1989) 93–99.

Glass and Glazing Federation, London. (1978): 'Glazing Manual'.

Gliemeroth, G. Eichorn, U. and Hölzel, E.: Zur Einflüssung der Eigenschaften silberhalogenhaltiger fototroper Gläser. Glastechn. Ber. **54** (1981) 162–174.

Goldner, R.B. and Rauh, R.D.: Electrochromic materials for controlled radiant energy transfer in buildings. Proc. SPIE **428** (1983) 38–44.

Gossink, R.G.: The preparation of ultra-pure glasses for optical waveguides J.Non-Cryst. Solids **26** (1977) 114–157.

Gottardi, V.(ed.): First International Workshop on Glasses and Ceramics from Gels. J. Non-Cryst. Solids **48** (1982).

Götz, H.; Helland, G.; Schaeffer, H.A.: Chemische Veränderung von Glasoberflächen durch Ionentransport in elektrischen Feld. Glastech. Ber. **52** (1979) 99–104.

Graham, P.W.L.: Lubricious and protective coatings for glass containers. Ceram. Engng. Sci. Prog. **7** (1986) no.3/4 439–59.

Graham, P.W.L. and Davey, R.G.: The processing of bottles and other hollow ware articles (Update). In: Tooley, F.V. (ed.) 'The Handbook of Glass Manufacture'. Vol II. 3rd Edn. Section 10. Ashlee Publishing Co. New York (1984).

Green, D.J.: Compressive surface strengthening of brittle materials. J. Mater. Sci. 19 (1984) 2165–2171.

Griscom, D.L.: The defect structure of glasses. J. Non-Cryst. Solids **73** (1985) 232–252

Griscom, D.L.: Borate glass structure. In: Pye,L.D., Fréchette, V.D. and Kreidl ,N.J.(eds) 'Borate Glasses'. Plenum Press, New York, London (1978a) 11–149.

Griscom, D.L.: Defects and impurities in _-quartz and fused silica. In: Pantelides, S.T. (ed) 'The physics of SiO_2 and its interfaces'. Pergamon. Oxford, New York (1978b) 232–252.

Grossmann, D.G.: Machinable glass-ceramics. J. Am. Ceram. Soc. **55** (1972) 446–449.

Groth, R. and Reichelt, W.: Gold-coated glass for the building industry. Gold Bulletin **7** (1974), no.3, 62–68.

Guillemet, C.: Current challenges concerning the mechanical strength of glass products. In: Kurkjian, C.J. (ed.) 'Strength of Inorganic Glass. NATO Conference Series. Series VI. Materials Science. Volume 11'. Plenum Press. New York (1985) 407–418.

Gulati, S.T. and Hagy, H.E.: Theory of the narrow sandwich seal. J. Am. Ceram. Soc. **61** (1978a) 260–263.

Gulati, S.T. and Hagy, H.E.: Finite element analysis and experimental verification of the shape factor for narrow sandwich seals. J. Am. Ceram. Soc. **61** (1978b) 263–267.

Gupta, P.K.: Effects of testing parameters on the tensile strengths of pristine E and S glass fibres. In: Kurkjian, C.J. (ed.) 'Strength of Inorganic Glass. NATO Conference Series. Series VI. Materials Science. Volume 11'. Plenum Press. New York (1985) 351–362.

Guy, A.G. and Newton, J.A.: Measurement of thermal performances of coated double glazed units. Proc IPAT workshop on window coatings for energy savings. Brussels (1984). Paper 19.

Hagy, H.E., Barney, W.H.; McCartney, J.S.; Parker, W.A. and Plummer, W.A. and Tooley, F.V.: Physical properties of glass. In: Tooley, F.V. (ed.) 'Handbook of Glass Manufacture. Vol. II 3rd Edn'. Ashlee Publishing Co. New York. (1984) 893–956.

Hagy, H.E. and Smith, A.F.: The sandwich seal in the development and control of sealing glasses. J. Canad. Ceram. Soc. **38** (1969) 63–68.

Hagy, H.E.: The trident seal. A rapid and accurate expansion differential test. J. Am. Ceram. Soc. **62** (1979) 60–62.

Hajto, J. and Owen, A.E.: Applications of amorphous semiconductor materials in electronics. Materials and Design **5** (Nov/Dec 1984) 221–227.

Hamilton, B. and Rawson, H.: Determination of flaw distributions on various glass surfaces by Hertz fracture experiments. J. Mech. Phys. Solids **18** (1970) 127–147.

Hamilton, B. and Rawson, H.: Flaw distributions in glass surfaces and mechanisms of damage. In: Douglas, R.W. and Ellis, B.(eds) 'Amorphous Materials'. Wiley-Interscience London (1972) 523–530.

Hamilton, B.: The incidence of finish crizzles during glass container manufacture. In: Proc. XIth International Congress on Glass. Prague (1977). Vol.5. 317–326.

Hammel, J.J.: Flat glass products J. Non-Cryst Solids **73** (1985) 363–378.

Hartig, K., Münz, W.D. and Scherer, M.: Industrial realisation of low-emittance oxide/metal/oxide films on glass. Proc. SPIE **428** (1983) 9–13.

Hauser, G: Spezialgasfüllung bei Mehrscheiben Isoliergläsern. Die Bedeutung für den Wärmeschutz. Glasswelt (1986) no.9. 42–53.

Hayward, P.J.: The use of glass-ceramics for immobilising high level wastes from nuclear fuel recycling. Glass Technol. **29** (1988) 122–136.

Heather, R.P. Forming machines. J. Non-Cryst. Solids **26** (1977) 337–371.

Hench, L.L.: Physical chemistry of glass surfaces. J. Non-Cryst. Solids **25** (1977) 343–369.

Hench, L.L.: Leaching of nuclear waste glasses. In: Wright, A.F.; and Dupuy, J. (eds) 'Glass... Current issues. NATO ASI series. Series E: Applied Sciences - No. 92'. Martinus Nijhoff. Dordrecht, Boston, Lancaster (1985a) 631–637.

Hench, L.L. and Spilman, D.B.: Composition and bonding mechanisms in 'Bioglass R' implants. In: Wright, A.F. and Dupuy, J. (eds) 'Glass... Current issues. NATO ASI series. Series E: Applied Sciences - No. 92'. Martinus Nijhoff. Dordrecht, Boston, Lancaster (1985b) 656–661.

Hendrickson, J.R. and Bray, P.J.: A theory for the mixed alkali effect. Parts 1 and 2. Physics Chem. Glasses **13** (1972) 43–49, 107–115.

Hewak, D.W. and Lit, J.W.Y.: Solution deposited optical waveguide lens. Appl. Opt. **28** (1989) 4190–4198.

Hiller, S.: Gas laser tubes: a challenge for glasses and ceramics. Glass Technol. **24** (1983) 302–308.

Hillig, W.B.: Strength of bulk fused quartz. J. Appl. Phys. **32** (1961) 741.

Hinz, P and Dislich, H.: Anti-reflecting light-scattering coatings produced via the sol–gel procedure J. Non-Cryst. Solids **82** (1986) 411–416.

Hlavac, J.: 'The technology of glass and ceramics'. Elsevier Scientific Publishing Co. Amsterdam, Oxford, New York. (1983) 237–243.

Höland, W., Vogel, W., Naumann, K. and Gummel, J.: Machinable bioactive glass-ceramic. In: Wright, A.F. and Dupuy, J. (eds) 'Glass... Current issues. NATO ASI series. Series E: Applied Sciences – No. 92'. Martinus Nijhoff, Dordrecht, Boston, Lancaster (1985) 670–675.

Höland, W., Vogel, W., Naumann, K. and Gummel, J.: Maschinell bearbeitbare bioaktive Glaskeramiken. Wiss. Zeitschr. der Friedrich-Schiller-Universit_t, Jena. Math.Nat.-Reihe **32** (1983) 571–580.

Holmberg, S.H., Shanks, R.R. and Bluhm, V.A.: Chalcogenide memory materials. J. Electronic Materials **8** (1979) 333–344.

Holscher, H.H.: The processing of bottles and other hollow ware articles. In: Tooley,F.V. (ed.) 'The Handbook of Glass Manufacture. Vol. II. 3rd Edn. Section 10'. Ashlee Publishing Co. New York.(1984).

Hood, H.P.; Nordberg, M.E.: US Patent 2 106 744 (1934) Treated borosilicate glass.

Hoogendorn, H. and Sunners, B.: IR absorbing sealing glasses. Am. Ceram. Soc. Bull. **48** (1969) 1125–1127.

Hooper, J.A.: On the bending of architectural laminated glass. Int. J. mech. Sci. **15** (1973) 309–323.

Horne, D.F.: 'Spectacle lens technology'. Adam Hilger. Bristol (1978).

Horne, D.F.: 'Optical instruments and their applications'. Adam Hilger. Bristol (1980).

Horne, D.F.: 'Optical production technology. 2nd Edn'. Adam Hilger. Bristol (1983).

Huber, P. and Gultmann, J.: Reproducibility of optical and mechanical properties of fibres prepared by the modified chemical deposition process. Physics Chem. Glasses **21** (1980) 43–46.

Hughes, K. and Isard, J.O.: Ionic transport in glass. In: Hladik, J.H. (ed.) 'Physics of Electrolytes Vol.1' Academic Press. London. (1972) 351–400.

Hull, A.W. and Burger, E.E.: Glass to metal seals. Physics **5** (1934) 387–405.

Hull, A.W.: Stresses in cylindrical glass–metal seals with glass inside. J. Appl. Phys. **17** (1946) 685–687.

Hunt, J.T., Johnson, B.C. and Kuizenga, D.J.: The Omega high power phosphate glass system: design and performance. J. Quant. Electronics **QE-17** (1981) 1620–1628.

Hussmann, E.: Gläser mit oberflächenschichten. Schott Information (1975). no 1, 1–6.

Hutchins, J.R. and Harrington, R.V.: Glass. In: 'Kirk-Othmer Encyclopaedia of Chemical Technology. 2nd Edn. Volume 10'. John Wiley New York. (1966) 533–604.

Ingram, M.D: Ionic conductivity in glass. Physics Chem. Glasses **28** (1987) 215–234

Ingram, M.D.: The impact of recent developments on the theory of and prospects for ionic conduction in glass. J. Non-Cryst Solids **73** (1-3) (1985) 247–254.

Isard, J.O.: Electrical conduction in the aluminosilicate glasses. J.Soc. Glass Technol. **43** (1959) 113–123.

Isard, J.O.: The mixed alkali effect in glass J. Non-Cryst. Solids **1** (1968/9) 235–261.

ISO 695 (1975): Mixed alkali test (or BS 3473 Part 1. 1982)

ISO 3825 (1977): Glass transfusion bottles for medical use – chemical resistance.

ISO 719 (1981): Grain test at 98 °C (or BS 3473 Part 2. 1982)

ISO 720 (1981): Grain test at 121 °C (or BS 3473 Part3. 1982)

ISO 4802 (1982): Whole article test (or BS 3473. Part 4. 1983)

ISO/DIS 9050: Glass in building. Determination of light transmission, direct solar transmittance, total solar energy transmittance, ultraviolet transmittance and related glazing factors.

Jackson, G.K.: The Pilkington low energy house at Milton Keynes. Glass Technol. **28** (1987) 124–128.

Jackson, J.D.J.; Rand, B.; Rawson, H.: Glass surface coatings resistant to mechanical damage. Thin Solid Films. **77** (1981) 5–12.

Jackson, N.; Ford, J.: Experience in the control and evaluation of coatings on glass containers. Thin Solid Films. **77** (1981) 23–39.

Janderleit, O.: Glass improvement by combination with plastics. Glastech. Ber. **62** (1989) 204–207.

Janowski, J. and Heyer, W.: 'Poröse Gläser. Herstellung, Eigenschaften und Anwendung'. VEB Deutscher Verlag für Grundstoffindustrie, Leipzig (1982)

Jarzebski, Z.M. and Marton, J.P.: Physical properties of SnO_2 materials. II Electrical properties. J.electrochem.Soc. **123** (1976a) 299C–332C.

Jarzebski, Z.M. and Marton, J.P.: Physical properties of SnO_2 materials. III Optical properties. J.electrochem.Soc. **123** (1976b) 333C–346C.

Jennings, R. and Wilberforce, R.R.: Thermal comfort and space insulation. Insulation. March 1973. 57–60.

Jiang Yasi, Zhang Junzhou, Xu Wenjiun *et al:* Preparation techniques for phosphate laser glasses. J. Non-Cryst. Solids **80** (1986) 623–629.

Johnston, W.D.: Oxidation-reduction equilibria in iron-containing glasses. J. Am. Ceram. Soc. **47** (1964) 198–201.

Jones, M.W. and Kao, K.C.: Spectrophotometric studies of ultra low loss optical glasses. II Double beam method. J.Sci.Instrumen. (J.Phys.E) **2** (1969) 331–335.

Jones, R.W.: Glass-ceramic coated substrates. Hybrid Circuits **6** (1985) 60–61.

Jones, R.W.: 'Fundamental principles of sol-gel technology'. Institute of Metals, London (1990) 136 pp.

Jones, S.P. and Williams, J.H.: Computer modelling as an aid to the thermal design of moulds for glass container manufacture. Glastechn.Ber. **56K** (1983) Vol.1 277–282.

Kaiser, A. and Schmidt, H.: Generation of SiO_2 membranes from alkoxysilanes on porous supports. J. Non-Cryst. Solids **63** (1/2) (1984) 261-271.

Kalbskopf, R.: Continuous chemical vapour deposition of tin oxide. Thin Solid Films **77** (1981) 65-67.

Kao, K.C.; Hockham, G.A.: Dielectric-fibre surface waveguides for optical frequencies. Proc. IEE **113** (1966) 1151–1158.

Kapron, F.P.; Keck, D.B. and Maurer, R.D.: Radiation losses in glass optical waveguides. Appl. Phys. Lett. **17** (1970) 423–425.

Katsuyama,T., Matsumura, H. and Suganuma, T.: European Patent Application 82301088.9 Publication No. 0 060 085 (1982)

Kawabata, T. *et al.*: Polyurethane emulsions for protective coatings of glass bottles. J. Elast. Plast. **13** (1981) 26–36.

Kay, S.E.: Glass and the car. Chemy. Ind. **23** (1973) 1086–94.

Keck, D.B., Maurer, R.D.; Schultz, P.C.: On the ultimate lower limit of attenuation in glass optical waveguides. Appl. Phys. Lett. **222** (1973) 307–308.

Keefer, K.D.: The effect of hydrolysis conditions on the structure and growth of silicate polymers. In: Brinker, C.J, Clark, D.E.and Ulrich D.R. (eds) 'Better Ceramics through Chemistry' North Holland, Amsterdam (1984) 15–24.

Kerkhof, F.: 'Bruchvorgänge in Gläsern'. Deutsche Glastechnische Gesellschaft, Frankfurt/Main (1970).

Kerper, M.J. and Scuderi, T.G.: Comparison of single point and two point bending for determining the strength of flat glass. Proc. Am. Soc. Test. Mater. **64** (1964) 1037–1043.

Kieffer, W. and Lindig, O.: Method for prestressing of glass. In: Kurkjian, C.J. (ed.) 'Strength of Inorganic Glass. NATO Conference Series. Series VI. Materials Science. Volume 11'. Plenum Press. New York (1985) 501–524.

King, R.D. and Wright, R.W.: Glass in high speed transport. Glass Technol. **28** (1987) 73–81.

Kishimoto, K., Aoki, S. and Sakata, M.: Effects of stress distribution on crack-branching in glass. Arch. Mech. **33** (1981) 947–956.

Klein, L.C. (ed.): 'Sol–gel technology for thin films, fibers, preforms and speciality shapes'. Noyes Publications (1988a)

Klein, L.C.: Filters and membranes by the sol-gel process. In: Klein, L.C. (ed.): 'Sol-gel technology for thin films, fibers, preforms and speciality shapes' Noyes Publications (1988b) 382–399

Klein, L.C. and Garvey, G.J.: Effect of water on acid– and base– catalysed hydrolysis of tetraethylorthosilicate. In: Brinker, C.J, Clark, D.E. and Ulrich D.R. (eds) 'Better Ceramics through Chemistry' North Holland, Amsterdam (1984) 33–39.

Klein, W.: Sonnenreflexionsglas für den Hochbau. Schott Information (1975). no 1, 8–13.

Klement, W., Willens, R.H. and Duwez, P.: Non-crystalline structure in solidified gold-silicon alloy. Nature, Lond. **187** (1960) 869–870.

Knott, P.: Glasses – Agricultural applications. Glastech. Ber. **62** (1989) 29–34.

Kohl, W.H.: Handbook of materials and techniques for vacuum devices. Reinhold, New York (1967).

Kolomiets, B.T.: Vitreous semiconductors. Phys. Stat. Sol. **7** (1964) 359–372; 713–731.

Köster, U.: Influence of solidification parameters and relaxation on properties of metallic glasses. In: Wright, A.F. and Dupuy, J. (eds) 'Glass... Current issues. NATO ASI series. Series E: Applied Sciences – No. 92'. Martinus Nijhoff. Dordrecht, Boston, Lancaster (1985) 79–85.

Köstlin, H., Jost, R. and Lems, W.: Optical and electrical properties of doped In_2O_3 films. Phys. Stat. Sol.(a) **29** (1975) 87–93.

Kreidl, N.J.: Inorganic glass-forming systems. In: Uhlmann D.R.and Kreidl N.J. (eds) 'Glass. Science and Technology.Volume 1. Glass-forming Systems'. Academic Press. London, New York. (1983) 105–299.

Kreidl, N.J. and Rood, J.L.: Optical materials. In: Kingslake, R. (ed) 'Applied Optics and Optical Engineering. Vol.1' Academic Press. San Francisco, New York, London. (1965) 153–200.

Kreidl, N.J.: Optical properties. In: Tooley, F.V. (ed.) 'Handbook of Glass Manufacture. Vol II. 3rd Edn'. Ashlee Publishing Co Inc. New York (1984) 957–997; 998/1–998/14.

Kuppers, D., Koenings, J. and Wilson, H.: Deposition of fluorine-doped silica layers from a $SiCl_4$/ SiF_4/ O_2 gas mixture by the plasma CVD method. J. Electrochem. Soc. **125** (1978) 1298–1302.

Kurkjian, C.J. (ed.) 'Strength of Inorganic Glass. NATO Conference Series. Series VI. Materials Science. Volume 11'. Plenum Press. New York (1985).

LaCourse, W.C.; Varshneya, A.K. and Alderson, D.: Containers for the 21st century: opportunities for lightweighting. J. Non-Cryst. Solids **73** (1985) 389–394.

La Course, W. C.: Continuous filament fibers by the sol-gel process. In: Klein, L.C. (ed.) 'Sol–gel technology for thin films, fibers, preforms and speciality shapes' Noyes Publications (1988) 184–198.

Lakatos, T., Johansson, L.-G. and Simminsköld, B.: Viscosity temperature relationships in the glass system SiO_2–Al_2O_3–Na_2O–K_2O–CaO–MgO in the composition range of technical glasses. Glass Technol. **13** (1972) 88–95.

Lampert, C.M.: Heat mirror coatings for energy conserving windows. Solar energy materials. **6** (1981) 1–41.

Larner, L.J., Speakman, K. and Majumdar, A.J.: Chemical interactions between glass fibres and cement. J. Non-Cryst. Solids **20** (1976) 43–74.

Laude, F.L.: French experience in vitrification of radioactive waste solutions. In: Wright, A.F. and Dupuy, J. (eds). 'Glass... Current issues. NATO ASI series. Series E: Applied Sciences –No. 92'. Martinus Nijhoff. Dordrecht, Boston, Lancaster (1985) 638–655

Lawson, K.R.: Contributions and effects of coatings on optical fibres. Proc SPIE **404** (1984) 109–118.

Le Comber, P.G.: Applications of amorphous silicon. In: Wright, A.F. and Dupuy, J. (eds) 'Glass... Current issues. NATO ASI series. Series E: Applied Sciences –No. 92'. Martinus Nijhoff, Dordrecht, Boston, Lancaster (1985) 430–448.

Levin, E.M. and Block, S.: Structural interpretation of immiscibility in oxide systems. I. Analysis and calculation of immiscibility. J. Am. Ceram. Soc. **40** (1957) 95–106.

Levine, C.A.: Recent progress in the development of the Dow hollow fibre sodium-sulphur battery. In: Proc. 16th Intersociety Energy Conversion Engng. Conference – Technologies for the Transition Vol. 1 – Atlanta 1981. ASME. New York (1981). 823–825

Lezal, D. and Srb, I.: Preparation of high purity chalcogenide glasses. Proc XIth Inter. Congr. Glass, Prague (1977). Vol 5. 497–506.

Lian Tiejun, Jiang Chuansong, Yan Guofeng, Xu Yueshen and Wang Naizhen: Sealing glasses with high absorption in the near infrared. J. Non-Cryst. Solids **52** (1982) 467–477.

Lind, N.C.: Approximate strength analysis for glass plates. J. Struct. Engng. ASCE **112** (1986) 1704–1720.

Lindig, O. and Pannhorst, W.: Thermal expansion and length stability of 'Zerodur' in dependence on temperature and time. Appl. Opt. **24** (1985) 3330–3334.

Linsley, G.S., Owen, A.E. and Hayatee, F.M.: Electronic conduction in vanadium phosphate glasses. J. Non-Cryst. Solids **4** (1970) 208–219.

Livage, J.: Transition metal oxide glasses. In: Wright, A.F. and Dupuy, J. (eds) 'Glass... Current issues. NATO ASI series. Series E: Applied Sciences – No. 92'. Martinus Nijhoff. Dordrecht, Boston, Lancaster (1985) 408–418.

Loewenstein, K.L.: 'The manufacturing technology of continuous glass fibres'. Elsevier. Amsterdam, New York. (1973).

Lomax, J.J.S.: The mechanical strength of glass containers. Glass Technol. **24** (1983) 27–39.

Luborsky, F.E. In: Levy, R.A. and Hasegawa, R. 'Amorphous Magnetism II'. Plenum Press. New York. (1977) 345–368.

Luborsky, F.E., Frischmann, P.G. and Johnson, L.A.: 'The role of amorphous materials in the magnetics industry'. J. Magnetism and Magnetic Materials **8** (1978) 318–325.

Luborsky, F.E.: Real and potential applications of amorphous metal ribbons In: Wright, A.F. Dupuy, J. (eds) 'Glass... Current issues. NATO ASI series. Series E: Applied Sciences – No. 92'. Martinus Nijhoff. Dordrecht, Boston, Lancaster (1985) 139–152.

Mackenzie, J.D.: Applications of the sol-gel process. J. Non-Cryst. Solids **100** (1988) 162–168.

McMaster, H.A.: Annealing and Tempering. In: Tooley, F.V. (ed.) 'Handbook of Glass Manufacture. Vol II 3rd Edn'. Ashlee Publishing Co Inc. New York (1984).

McMillan, P.W.: 'Glass-Ceramics. 2nd Edition'. Academic Press. London, New York, San Fransisco (1979).

McMillan, P.W., Hodgson, B.P. and Partridge, G.: Sealing glass-ceramics to metals. Part 1. Selection of materials and direct sealing methods. Glass Technol. **7** (1966a) 121–127.

McMillan, P.W., Partridge, G , Hodgson, B.P. and Heap, H.R.: Sealing glass–ceramics to metals. Part 2. Sealing by intermediate bonds. Glass Technol. **7** (1966b) 128–133.

Majumdar, A.J.: Glass fibre reinforced cement and gypsum products. Proc. R. Soc. **A 319** (1970) 69–78.

Manfre, G.: Strength of automotive window glass. In: Kurkjian, C.J. (ed.) 'Strength of Inorganic Glass. NATO Conference Series. Series VI. Materials Science. Volume 11'. Plenum Press. New York (1985) 423–428.

Manifacier, J.C., Fillard, J.P.and Bind, J.M.: Deposition of In_2O_3 – SnO_2 layers on glass substrates using a spraying method. Thin Solid Films 77 (1981) 67–80.

Marsh, J.A.: The airborne sound insulation of glass. Applied Acoustics. Part 1. **4** (1971a) 55–70.

Marsh J.A.: ibid. **4** (1971b) 131–154.

Marsh J.A.: ibid. **4** (1971c) 175–191.

Maruyama, E., Hirai,T., Goto, N. *et al.*: Photoelectric properties of chalcogenide glass diodes and their application to TV pick up tubes. Proc. 5th Int. Congr. Amorphous and Liquid Semiconductors. Taylor and Francis. London (1973) 581–594.

Maurer, R.D.: Nucleation and growth in a photosensitive glass. J. Appl. Phys. **29** (1958) 1–8.

Maurer, R.D.: Doped-deposited silica fibres for communications. Proc. IEE. **123**(1976) 581–585.

Maurer, R.D.: Glass fibres as optical waveguides. J. Non-Cryst. Solids 25 (1977) 324–342.

Mazurin, O.V., Streltsina, M.V.and Shvaiko-Shvaikovskaya, T.P.: Handbook of Glass Data. Part A. Silica glass and binary silicate glasses. Elsevier. Amsterdam, Oxford, New York (1983).

Mazurin, O.V., Streltsina, M.V.and Shvaiko-Shvaikovskaya, T.P.: Handbook of Glass Data. Part B. Single component and binary non-silicate oxide glasses Elsevier. Amsterdam, Oxford, New York (1985).

Mazurin, O.V., Streltsina, M.V.and Shvaiko-Shvaikovskaya, T.P.: Handbook of Glass Data. Part C. Ternary silicate glasses. Elsevier. Amsterdam, Oxford, New York (1987).

Mecholsky, J.J., Freiman, S.W.and Rice, R.W. Fracture surface analysis of ceramics. J. Mater. Sci, **11** (1976) 1310–1319.

Megla, G.K.: Optical properties and applications of photochromic glass. Appl. Opt. **5** (1966) 945–960.

Meinecke, G.: Über die Austauchbarkeit thoroxydhaltiger Gläser für die Optik. Glas-Email-Keramo-Technik **10** (1959) 209–212.

Meistring, R., Frischat, G.H.and Hennicke, H.W.: Kinetische Vorgänge beim Farbbeizen von Gläsern. Glastech. Ber. **49** (1976) 60–66.

Messing, R.A.: Glass as a bioactive material. J. Non-Cryst. Solids 26 (1977) 483–513.

Mettes, D.G.: Glass fibers. In: Lubin, G. (ed.) 'Handbook of Fiberglass and Advanced Plastics Composites'. Van Nostrand-Reinhold. New York, London (1969) 143–190.

Minami, T.: Fast ion conducting glasses. J.Non-Cryst Solids **73** (1–3) (1985) 273–284.

Minor, J.E.: Window glass design practice: A review. J Structur. Div. ASCE **107** (1981) 1–12.

Miya ,T., Terunuma, Y., Hosaka, T. *et al.*: Ultimate low loss single mode fibres at 1.55 μm. Electron. Lett. **15** (1979) 106–108

Mohr, J.G. *et al.* (eds): 'SPI Handbook of Technology and Engineering of Reinforced Plastics/Composites'. 2nd Edn. (1973). New York. Van Nostrand.

Mohr, J.G.and Rowe, W.P.: 'Fiber Glass'. Van Nostrand-Reinhold. New York, London (1978).

Moody, B.: 'Packaging in Glass'. Hutchinson-Benham. London (1977).

Moriya, Y., Warrington, D.H. and Douglas, R.W.: A study of the metastable liquid-liquid immiscibility in some binary and ternary silicate glasses. Physics Chem. Glasses **8** (1967) 19–25.

Mott, N.F.and Davis, E.A.: 'Electronic processes in non-crystalline solids. 2nd Edn. Clarendon Press', Oxford (1979).

Mukkerjee, S.P. and Philippou, J.: Glassy thin films and fibrisation by the gel route. In: Wright, A.F.and Dupuy, J. (eds) 'Glass... Current issues. NATO ASI series. Series E: Applied Sciences - No. 92'. Martinus Nijhoff. Dordrecht, Boston, Lancaster (1985) 232-253.

Müller-Simon, H. and Barklage-Hilgefort, H.: Strength optimisation of glass containers by the finite element method. Glastech. Ber. **61** (1988) 348–357.

Murgatroyd, J.B.: The significance of surface marks on fractured glass. J. Soc. Glass Technol. **26** (1942) 155–171T.

Nassau, K. and Shiever, J.W.: Plasma torch preparation of high-purity low OH content fused silica. Am. Ceram. Soc. Bull. 54 (1975) 1004–1009, 1011.

Nath, P.and Douglas, R.W.: Cr^{3+}-Cr^{6+} equilibrium in binary silicate glasses. Physics Chem. Glasses **6** (1965) 197–203.

Neale, R.G.: Device structures and fabrication techniques for amorphous semiconductor switching devices. J. Non-Cryst. Solids **2** (1970) 558–574.

Nebrensky, J.: Photosensitive glasses and their properties. In: 'Symposium on Coloured Glasses. Jablonec'. Czechoslovak Scientific Society Prague (1965) 173–181.

Nelson, D.L.: Ovonic device applications. J. Non-Cryst. Solids **2** (1970) 528–539.

Neuroth, N.: Laser glass: status and prospects. Opt. Engng. **26** (1987) 96–101.

Newton, R.G.: The durability of glass – a review. Glass Technol. **26** (1985) 21–38.

Newton, R.G.: 'The deterioration and conservation of painted glass: a critical bibliography'. British Association and Oxford University Press (1982).

Nicoletti, F., Geotti-Bianchini, F. and Polato, P.: Spectrophotometric determination of the normal emissivity of coated flat glass. Glastech. Ber. **61** (1988) 127–139.

Nordberg, M.E. and Hood, H.P.: US Patent 2 106 744 (1934).

Notis, M.R.: Decarburisation of an iron–nickel–cobalt alloy. J. Amer. Ceram. Soc. **45** (1962) 412–416.

Ohta, H.: The strengthening of mixed alkali glass by ion exchange. Glass Technol. **16** (1975) 25–29.

Oldfield, L.F.: The effects of heat treatments on the thermal expansion coefficients of some borosilicate glasses and iron–nickel–cobalt alloys and their significance in the behaviour of glass-to-metal seals. Glastech. Ber **32K** (1959) no V. 16–25.

Oliver, D.S.: Glass for construction purposes. J. Non-Cryst Solids **26** (1977) 515-602.

Ono, H.: Production of chemically strengthened bottles in the Japanese market. Glass Technol. **22** (1981) 173–181.

Orcel, G. and Hench, L. L.: Physical–chemical variables in processing $Na_2O-B_2O_3-SiO_2$ gel monoliths. In: Brinker, C.J., Clark, D.E. and Ulrich, D.R. (eds) 'Better Ceramics through Chemistry' North Holland. Amsterdam (1984) 79–84.

Orr, L.: Practical analysis of fracture of glass windows. Materials Res. Standards **12** (1972) 21–23, 47.

Ovshinsky, S.R.: Reversible switching phenomena in disordered structures Phys. Rev. Letters **21** (1968) 1450–1453.

Owen, A.E. and Douglas, R.W.: The electrical properties of vitreous silica. J. Soc. Glass Technol. **43** (1959) 159–78.

Owen, A.E.: Properties of glasses in the system $CaO-Al_2O_3-B_2O_3$. Part I. The dc conductivity and structure of calcium aluminoborate glasses. Physics Chem. Glasses **2** (1961) 87–98.

Owen, A.E.: Electric conduction and dielectric relaxation in glass. In: Burke, J.E.(ed.) 'Progress in Ceramic Science. Vol. 36. Pergamon. Oxford, New York (1963)

Owen, A.E.: Semiconducting glasses. I. Glass as an electronic conductor. Contemp. Phys. **11** (1970a) 227–255.

Owen, A.E.: Semiconducting glasses. II. Properties and interpretation. Contemp. Phys. **11** (1970b) 257–286.

Owen, A.E.: The electrical properties of glasses. J. Non-Cryst. Solids. **25** (1977) 370–423.

Owen, A.E., Robertson, J.M. and Main, C.: The threshold characteristics of chalcogenide glass memory switches. J. Non-Cryst Solids **32** (1979) 29–52.

Owen, A.E.: Electronic transport in amorphous chalcogenide semiconductors .In: Wright, A.F.; Dupuy, J. (eds) 'Glass... Current issues. NATO ASI series. Series E: Applied Sciences–No. 92'. Martinus Nijhoff. Dordrecht, Boston, Lancaster (1985) 376–398.

Owens, D.R.J., Williams, J.H. and Sa, J.M.A.C.: The numerical modelling of glass forming processes. Proc. XIV Intern. Congr. on Glass, New Delhi (1986). Vol.3. 138–145.

Owens, P.G.T.: Effective U values. Building Services Research and Technology **3** (1982) 189–192.

Pantano, C.G., Clark, A.E. and Hench, L.L.: Multilayer corrosion films on 'Bioglass R' surfaces. J. Am. Ceram. Soc. **57** (1974) 412–413.

Parker, C.J.: Optical materials – refractive. In: Shannon, R.R. and Wyant, J.C. (eds) 'Applied Optics and Optical Engineering. Vol.7'. Academic Press. San Francisco, New York, London. (1979) 47–77.

Partridge, G. and Phillips, S.V.: New developments in glasses and ceramics for the electrical industry. GEC J. Sci. Technol. **45** (1978) 11–20.

Partridge, J.H.: 'Glass–to–metal seals'. Society of Glass Technology Sheffield (1949).

Paschke, H.: Switching without touching. Schott Information (1978) Part 4 15–18.

Patek, K.: 'Glass lasers'. Illiffe. London (1970).

Paul, A. and Douglas, R.W.: Optical absorption of divalent cobalt in binary alkali borate glasses and its relation to the basicity of glass. Physics Chem. Glasses **9** (1968) 21–26.

Paul, A.: 'Chemistry of glasses'. Chapman and Hall. London, New York (1982).

Pederson, L.R.: Comparison of stannous and stannic chloride as sensitising agents in the electroless deposition of silver on glass using X-ray photoelectron spectroscopy. Solar Energy Materials. **6** (1982) 221–32.

Pederson, L.R. and Thomas, M.T.: Characterisation of new and degraded mirrors with AES, ESCA and SIMS. Solar Energy Materials **3** (1980) 151–167.

Persson, H.R.: Chemical durability of containers treated with various agents. Glass Technol. **3** (1962) 17–35.

Persson, R.: Flat glass technology. Butterworth. London (1969).

Petzold, A. and Marusch, H.: 'Der Baustoff Glas'. VEB Verlag der Bauwesen, Berlin (1973).

Philipp, G. and Schmidt, H.: New materials for contact lenses prepared from Si- and Ti_ alkoxides by the sol–gel process. J. Non-Cryst. Solids **63** (1/2) (1984) 283–292.

Phillips, S.V., Crozier, D.S., McMillan, P.W.and Taylor, J.McC.: Sea water desalination using glass hollow fibre membranes. Desalination **14** (1974) 209–216.

Physical Properties Committee of the Society of Glass Technology: Devitrification measurements. Glass Technol. **5** (1964) 82–87.

Pilkington, L.A.B.: The float process. Proc. R. Soc. **A 314** (1969) 1–25.

Pilkington, L.A.B.: Float – an application of science, analysis and judgement. Glass Technol. **12** (1971), 76–83.

Pilkington, L.A.B.: Flat glass – evolution and revolution over 60 years. Glass Technol. **17** (1976), 182–193.

Pilkington Bros Ltd: 'Glass and thermal safety'.

Pilkington Bros Ltd: 'Glass and noise control'.

Piscatelli, R.A., Rhee, S.K.and Bradley, R.N.: Oxidation of Fe–29Ni–17Co alloy. J. Electrochem Soc. **123** (1976) 929–933.

Plueddemann, E.P.: Int. J. Adhesion and Adhesives (October 1981) 305-310

Poole, J.P. and Snyder, H. C.: Chemically strengthening glass containers by ion exchange. Glass Technol. **16** (1975) 109–113.

Poritski, H.: Analysis of thermal stresses in sealed cylinders. Physics **5** (1934) 406–414.

Poulain, M., Poulain, M. and Lucas, J.: Verres fluores au tetrafluorure de zirconium. Properties optique d'un verre dope au Nd^{3+}. Mater. Res. Bull. **10** (1975)243–6.

Prassas, M, Phallipou, J.and Zarzycki, J.: Synthesis of monolithic silica gels by hypercritical solvent evacuation. J. Mater. Sci. **19** (1984) 1656–1665.

Preston, F.W.: The propagation of fissures in glass and other bodies with special reference to the split-wave front. J. Am. Ceram. Soc. **14** (1931) 419–427.

Preston, F.W.: Bottle breakage – causes and types of fracture. Am. Ceram. Soc. Bull. **18** (1939) 35–60.

Proctor, B.A.: The long term behaviour of glass fibre reinforced composites. In: Wright, A.F.and Dupuy, J. (eds). 'Glass... Current issues. NATO ASI series. Series E: Applied Sciences – No. 92'. Martinus Nijhoff. Dordrecht, Boston, Lancaster (1985a) 524–550.

Proctor, B.A. and Yale, B.: Glass fibres for cement reinforcement. Phil. Trans. R. Soc. London **A294** (1980) 427–436.

Proctor, B.A.: Alkali resistant fibre for reinforcement of cement. In: Wright, A.F.andDupuy, J. (eds) 'Glass... Current issues. NATO ASI series. Series E: Applied Sciences – No. 92'. Martinus Nijhoff. Dordrecht, Boston, Lancaster (1985b) 555–573

Pulker, H.K.: 'Coatings on glass'. Elsevier. Amsterdam, Oxford, New York, Tokyo (1984).

Puyané, R. and Kato, I.: Tin oxide films on glass substrates by a sol–gel technique. SPIE Seminar No.401 (1983) Thin film technologies.

Rabinovich, E.M.: Crystallisation and thermal expansion of solder glass in the $PbO–ZnO–B_2O_3$ system with admixtures. Am. Ceram. Soc. Bull. **58** (1979) 595–8, 605.

Rabinovich, E.M.: Particulate silica gels and glasses from the sol–gel process. In: Klein, L.C. (ed.) 'Sol-gel technology for thin films, fibers, preforms and speciality shapes'. Noyes Publications (1988) 260–294.

Ramaswamy, R.V. and Srivastave, R.: Ion-exchanged glass waveguides: A review. J. Lightwave Technol. **6** (1988) 984–1000.

Ravaine, D.: Ionic conductive glasses. In: Wright, A.F.andDupuy, J. (eds) 'Glass... Current issues. NATO ASI series. Series E: Applied Sciences – No. 92'. Martinus Nijhoff. Dordrecht, Boston, Lancaster (1985) 435–455.

Rawson, H.: The theory of stresses in two-component glass–to–metal tube seals. J. Sci. Instr. **26** (1949) 25–27.

Rawson, H.: A theory of stresses in glass butt seals. Brit. J. Appl. Phys. **2** (1951) 151–6.

Rawson, H.: The relationship between liquidus temperature, bond strength and glass formation. Proc. IVth Internat. Congr. on Glass, Paris. (1956) 62–69.

Rawson, H: The calculation of transmission curves of glass stained with copper and silver compounds. Physics Chem. Glasses **6** (1965) 81–84.

Rawson, H.: 'Inorganic glass-forming systems'. Academic Press. London, New York. (1967).

Rawson, H.: 'Properties and applications of glass'. Elsevier. Amsterdam, Oxford, New York (1980).

Rawson, H.: Some aspects of the performance of low emissivity glass. Riv. Stn. Sper. Vetro **16** (1986) 137–40.

Rawson, H. and Giotti-Bianchini, F.: Damage resistant coatings on glass containers. In: Kurkjian, C.J. (ed.) 'Strength of Inorganic Glass. NATO Conference Series. Series VI. Materials Science. Volume 11'. Plenum Press. New York (1985) 453–468.

Rawson, H.: Why do we make glass so weak? A review of research on damage mechanisms. Glastech. Ber. **61** (1988)231–246.

Rawson, H.: Dependence of emissivity on angle for coated and uncoated glass and the calculation of radiant energy exchange in double glazed units. Glastech. Ber. **62** (1989) 167–174.

Ray, N.H.: 'Inorganic polymers'. Academic Press. London, New York (1978).

R.C.A.: Special issue on porcelain-enamelled-steel boards for electronic applications. RCA Review **42** (June 1981).

Rekhson, S.M.and Mazurin, O.V.: Stress relaxation in glass and glass to metal seals. Glass Technol **18** (1977) 7–14.

Rekhson, S.M.: Annealing of glass to metal and glass to ceramic seals. Part 1 Theory. Glass Technol. **20** (1979a) 27–35.

Rekhson, S. M.: Annealing of glass to metal and glass to ceramic seals. Part 2. Experimental. Glass Technol. **20** (1979b) 132–143.

Renfrew, C.: 'Before civilization'. Penguin Books. London (1976).

Roberts, G.J.: $FeO–K_2O–P_2O_5$ glasses as a source of micronutrient iron in the soil. Amer. Ceram. Soc. Bull. **54** (1975) 1069–1071.

Roeder, E. and Hilpert, H.-G.: Zum Verformungsverhalten einer Alkali–Kalksilicatglasschmelze in der Nähe der Transformationstemperatur. Festigkeitssteigerung durch Warmverformung. Glastech. Ber. **55** (1982) 96–101.

Ross, I.N., White, M.S.and Boon, J.E.: Vulcan – a versatile high power glass laser for multiuser experiments. J. Quant. Electronics **QE-17** (1981) 1653–1661.

Rubin, M.: Calculating heat transfer through windows. Energy Research **6** (1982a) 341–9.

Rubin, M.: Solar optical properties of windows. Energy Research **6** (1982b) 123–133.

Saeki, K. and Yamada, T.: Studies of solid film lubricants as substitutes for mould dope in glass bottle forming process. Journal JSLE. International Edition. (1982) no.3 91–94.

Sakka, S.: Fibers from the sol–gel process. In: Klein, L.C. (ed.)'Sol–gel technology for thin films, fibers, preforms and speciality shapes'. Noyes Publications (1988a) 140–161.

Sakka, S. (ed.): 'Fourth International Workshop on Glasses and Ceramics from Gels' J. Non-Cryst. Solids **100** (1988b).

Sarkisov, P.D.: The modern state of technology and application of glass - ceramics. Proc. XVth International Congress on Glass. Leningrad. (1989).

Savage, J.A. and Nielsen, S.: Chalcogenide glasses transmitting in the infrared between 1 and 20 μ - a state of the art review. Infrared Physics **5** (1965) 195–204.

Savage, J.A., Webber, P.J. and Pitt, A.M.: The potential of Ge–As–Se–Te glasses as 3–5 μm and 8–12 μm infrared optical materials. Infrared Physics **20** (1980) 313–320.

Savage, J.A.: Chalcogenide glasses for optical applications. In: Wright, A.F.and Dupuy, J. (eds). 'Glass... Current issues. NATO ASI series. Series E: Applied Sciences – No. 92'. Martinus Nijhoff. Dordrecht, Boston, Lancaster (1985) 281–306

Schaeffer, H.A.: Thermal and chemical strengthening of glass – Review and outlook. In: Kurkjian, C.J. (ed) 'Strength of Inorganic Glass. NATO Conference Series. Series VI. Materials Science. Volume 11'. Plenum Press. New York (1985) 469–483.

Schairer, J.F. and Bowen, N.L.: The system $Na_2O–Al_2O_3–SiO_2$. Amer. J. Sci. **254** (1956) 129–195.

Schauer, A.: The dawning of the flat panel monitor. Schott Information (1982) Part 3. 16–19.

Scheidler, H.: Beheizen von Ceran Kochflachen. Schott Information. Part 1 (1974) 1–18.

Scheidler, H. and Kristen, K.: Ceran: the perfect cooking system. Schott Information. Part 1 (1980) 2–17.

Scheidler, H. and Taplan, M.: Ceran glass ceramic hob tops. Schott Information Part 2 (1984) 2–21.

Scherer, G.W. and Schultz, P.C.: Unusual methods of producing glasses. In: Uhlmann, D.R.and Kreidl, N.J. (eds.) 'Glass: Science and technology. Vol.1. Glass-forming systems'. Academic Press. New York, London, Paris. (1983) 49–103.

Scherer, G.W.: Glasses and ceramics from colloids. In: Brinker, C.J., Clark, D.E. and Ulrich, D.R. (eds). 'Better Ceramics through Chemistry'. North Holland. Amsterdam (1984) 205–211.

Scherer, G.W. and Rekhson, S.M.: Viscoelastic analysis of stresses in composites. In: Tomozawa, M.; Doremus, R.H. (eds). 'Treatise on Materials Science and Technology. Volume 26. (Glass IV)'. Academic Press. New York, London (1985) 245–318.

Schnabel, H. and Langer, P.: Structural and chemical properties of glass capillary membranes and their use in protein separation. Glastech. Ber. **62** (1989) 56–62.

Schmidt, H.: Organically modified silicates by the sol-gel process. In: Brinker, C.J,Clark, D.E., Ulrich, D. R. (eds).'Better Ceramics through Chemistry'. North Holland. Amsterdam (1984) 327–336.

Schmidt H.: Films by sol–gel processes. In: Picozzi, P. *et al.* (eds) Proc. European Meeting on Inorganic Coatings on Glass. L'Aquila (1988). COMMET. 131–148.

Schneider, E.: Polymer coating of glass containers. Glastech. Ber. **50** (1977) 201–205.

Scholze, H.: 'Glas-Natur, Struktur und Eigenschaften. 3rd Edn'. Springer. New York, Heidelberg, Berlin (1988).

Scholze, H. (ed.) 'Second International Workshop on Glasses and Ceramics from Gels' J. Non-Cryst. Solids **63** (1/2) (1984).

Schroeder, H.: Oxide layers deposited from organic solutions. In: Hass, G.and Thun, R.E. (eds.) 'Physics of Thin Films Vol. 5'. Academic Press. New York.(1969) 87–140.

Schultz, P.C. and Smyth, H.T.: Ultra-low expansion glasses and their structure in the SiO_2–TiO_2 system. In: Douglas, R.W.and Ellis, B. (eds) 'Amorphous Materials'. Wiley Interscience. London, New York (1970) 445–461.

Schultz, P.C.: Recent advances in optical fibre materials. 2nd International Otto Schott Colloquium. Wiss. Zeitschr. der Friedrich-Schiller-Universität, Jena. Math. Nat.-Reihe **32** (1983) 215–226.

Schuster, E.: New dimensions in astronomy. Schott Information 48/88 4-9.

Schweig, B.: 'Mirrors'. Pelham Books. London (1973).

Scott, B. and Rawson, H.: Preparation of low loss glasses for optical fiber communication systems. Opto-electronics **5** (1973) 285–288.

Selkowitz, S.E.: Thermal performance of insulating window systems. ASHRAE Trans. **85** (1979) 669–685.

Shartsis, L., Capps, W. and Spinner, S.: Viscosity and electrical resistivity of molten alkali borates. J. Am. Ceram. Soc. **36** (1953) 319–326.

Shartsis, L., Capps, W. and Spinner, S.: Density, expansivity aand viscosity of molten alkali silicates. J. Am. Ceram. Soc. **35** (1952) 155–160.

Shelby, J.E., Vitko Jr, J. and Farrow, R.L.: Characterisation of heliostat corrosion. Solar Energy Materials **3** (1980) 185–201.

Sieger, J.S.: Chemical characteristics of float glass surfaces J. Non-cryst Solids **19** (1975) 213–220.

Sigel, G.H.: Optical absorption of glasses. In: Tomozawa, M.and Doremus, R.H. 'Treatise on Materials Science and Technology. Vol.12. Glass I. Interaction with Electromagnetic Radiation'. Academic Press. New York, San Francisco, London (1977) 5–89.

Simiu, E. and Hendrickson, E.M.: Design criteria for glass cladding subjected to wind loads. J. Struct. Engrg. ASCE. **113** (1987) 501–518.

Simpson, H.E.: Method of measuring the surface durability of glass. Am. Ceram. Soc. Bull. **30** (1951) 41–5.

Smay, G. L.: Interaction of organic coatings with metal oxide surfaces and glass. Glass Technol. **26** (1985) 46–59.

Smets, B.M.J. and Lommen, T.P.A.: The incorporation of aluminium oxide and boron oxide in sodium silicate glasses, studied by x-ray photoelectron spectroscopy. Physics Chem. Glasses **22** (1981) 158–162.

Smith, C.H.: Metallic glasses in high-energy pulsed power systems. In: Wright, A.F.and Dupuy, J. (eds) 'Glass... Current issues. NATO ASI series. Series E: Applied Sciences – No. 92'. Martinus Nijhoff. Dordrecht, Boston, Lancaster (1985) 188–199.

Smith, G.P.: Photochromic glasses: Properties and applications. J. Mater. Sci. **2** (1967) 139–152.

Smith, G.P.: Some light on glass. Glass Technol. **20** (1979) 149–157.

Snitzer, E.: Glass lasers. Appl. Opt. **5** (1966) 1487–1499.

Snitzer, E.: Lasers and glass technology. Am. Ceram. Soc. Bull. **52** (1973) 516–525.

Soules, Th.F.: Computer simulation of glass structure. Proc. XVth Internat. Congr. on Glass, Leningrad (1989) 84–102.

Southwick, R.D., Wasylyk, J.S. and Smay, G.L. *et al.*: The mechanical properties of films for the protection of glass surfaces. Thin Solid Films **77** (1981) 41–50.

Southwick, R.D. and Wasylyk, J.S.: Recent developments in quality control in the glass container industry. Glass **63** (1986) 'Glassman 86' C59–C63.

Sowman, H.G.: Alumina–boria–silica ceramic fibres from the sol–gel process. In: Klein, L.C. (ed.) 'Sol–gel technology for thin films, fibers, preforms and speciality shapes'. Noyes Publications (1988) 162–183.

Sozanski, M.R.; Varshneya, A.K.: Strengthening of glass tubes and containers by flame sprayed glazing. Am. Ceram. Soc. Bull. **66** (1987) 1630–1634.

Speck, D.R., Bliss, E.S. and Glaze, J.A.: The Shiva laser fusion facility J. Quant. Electronics **QE-17** (1981) 1599–1619.

Spoor, W.J. and Burggraaf, A.J.: The strengthening of glass by ion exchange. Part 3. Mathematical description of the stress relaxation after ion exchange in alkali aluminosilicate glasses. Physics Chem. Glasses **7** (1966) 173–177.

Stahn, D.: Wärmespannung in grossflächigen Vergläsungen. Glastech. Ber. **50** (1977) 149–158.

Stavrinidis, B.and Holloway, D.G.: Crack healing in glass. Physics Chem. Glasses **24** (1983) 19–25

Stewart, D.R. and Spear, B.W.: Developments in asbestos-free hot glass handling materials. Glass **61** (1984) No.3 90, 92.

Stokes, J.: Platinum in the glass industry. zirconia grain stabilised materials supplement conventional alloys. Glass Technol. **28** (1987) 218-222.

Stookey, S.D.: Photosensitive glass. Ind.Eng.Chem. **41** (1949) 856–861

Stookey, S.D., Beall, G.H. and Pierson, E.J.: Full colour photosensitive glass J. appl. Phys. **49** (1978) 5114–5123.

Strnad, Z.: 'Glass-ceramic materials'. Elsevier. Amsterdam, Oxford, New York (1986).

Stroud, J.S.: The strengthening of some commercial opthalmic glasses and filter glasses by ion exchange. Glass Technol. **29** (1988) 108–114.

Svensson, J.S.E.M. and Granqvist, G.C.: Electrochromic coatings for smart windows: Crystalline and amorphous WO_3 films. Thin Solid Films **126** (1985) 31–36.

Swift, H.R.: How surface chemistry affects float glass properties. Glass Industry **65** (1984) no.5 27–30.

Szczyrbowski, J., Dietrich, A. and Hartig, K.: Bendable silver-based low emissivity coating on glass. Proc. SPIE. **823** (1987) 21–27.

Takamori,T.: Solder glasses. In: Tomozawa, M.and Doremus, R.H. (eds) 'Treatise on Materials Science and Technology. Vol. 17. Glass II'. Academic Press. New York, London. (1979) 173–235.

Telfer, S.B.,Zervas, G. and Knott, P.: UK Patent 2 116 424 B. (1986).

Terai, R.; Hayami, R.: Ionic diffusion in glasses. J. Non-Cryst. Solids **18** (1975) 217–64.

Thomas Jr., T.M., Pitts, J.R. and Czanerna, A.W.: Surface analysis of commercially made mirrors. Applications of Surface Science. **15** (1983) 75–92.

Thomas, I.M.: Multicomponent glasses from the sol–gel process. In: Klein, L.C. (ed.) 'Sol–gel technology for thin films, fibers, preforms and speciality shapes'. Noyes Publications (1988) 2–15.

Tiwari, A.N. and Das, A.R.: Correlation of refractive index and density in oxide glasses of high refractive index. Am. Ceram. Soc. Bull. **51** (1972) 695–697.

Tiwari, A.N. and Das, A.R.: Reflex refractive characteristics of surfaces coated with glass microspheres. Am. Ceram. Soc. Bull. **52** (1973) 175–179.

Tooley, F.V. (ed.): 'Handbook of Glass Manufacture'. 3rd Edn. Ashlee Publishing Co Inc. New York (1984).

Tran, D.C., Sigel, G.H. and Bendow, B.: Heavy metal fluoride glasses: a review. J Lightwave Technol. **LT-2** (1984) 566–586.

Trap, H.L. and Stevels, J.M.: Physical properties of invert glasses. Glastechn. Ber. **32K** (1959) VI 31–52.

Trier, F.: New products and future developments of automotive glazing. In: Picozzi *et al.* (eds) Proceedings of the European Meeting on Inorganic Coatings on Glass. L'Aquila (1988) COMMET 269-291.

Tsai, C.R. and Stewart, R.A.: Stress analysis of large deflection of glass plates by the finite element method. J. Am. Ceram. Soc, **59** (1976) 445–448.

Tuller, H.L.and Barsoum, M.W.: Glass solid electrolytes: past, present and near future – the year 2004. J. Non-Cryst. Solids **73** (1–3) (1985) 331–350.

Tummala R.R. and Shaw, R.R.: Glasses in microelectronics in the information-processing industry. In: Boyd, D.C.and MacDowell J.F.(eds): 'Commercial Glasses' American Ceramic Society, Columbus OH (1986) 87–104.

Turner, D.P.T. (ed): 'Window glass design guide'. Architectural Press, London (1977).

Turner, W.E.S. and Winks, F.: The thermal expansion of glass. J. Soc. Glass Technol. **14** (1930) 84–126.

Uhlmann, D.R.: Glass formation. J.Non-Cryst. Solids **25** (1977) 42–85.

Uhlmann, D.R. and Kreidl, N.J. (eds) 'Glass. Science and Technology. Volume 1. Glass-forming Systems'. Academic Press. London, New York (1983a).

Uhlmann, D.R. and Yinnon, H.: The formation of glasses. In: Uhlmann D.R.and Kreidl N.J. (eds) 'Glass. Science and Technology.Volume 1. Glass-forming Systems'. Academic Press. London, New York (1983b). 1–48

Uhlmann, D.R.: Nucleation and crystallization in glass-forming systems. In: Wright, A.F.and Dupuy, J. (eds) 'Glass... Current issues. NATO ASI series. Series E: Applied Sciences – No. 92'. Martinus Nijhoff. Dordrecht, Boston, Lancaster (1985) 1–20.

Uhlmann, D.R., Tuller, H.L., Button, D.P. and Valez, M.: Solid electrolyte batteries and fast ion conducting glasses. Factors affecting a proposed merger. 2nd International Otto Schott Colloquium. Wiss. Zeitschr. der Friedrich-Schiller-Universit_t, Jena. Math.Nat.-Reihe **32** (1983) 285–302.

Ulrich, D.R.: Prospects of sol–gel processes. J.Non-Cryst. Solids **100** (1988) 174–193.

Vail, J.G.: 'Soluble silicates'. 2 vols. Reinhold. New York (1952).

Vallabhan, C.V.G.: Iterative analysis of nonlinear glass plates. J. Struct. Engrg. ASCE. **109** (1983) 489–502.

Vallabhan, C.V.G. and Chou, G.D.: Interactive nonlinear analysis of insulating glass units. J. Struct. Engng. ASCE **112** (1986) 1313–1326.

Vallabhan, C.V.G., Minor, J.E. and Nagalla, S.R.: Stresses in layered glass units and monolithic glass plates. J. Struct. Engrg. ASCE **113** (1987) 36–43.

Valstar, P.: Pressing large colour TV tubes to close dimensional tolerances. Glass Technol. **20** (1979) 252–256.

Van der Sande, J.B. and Freed, R.L.: Metallic glasses. In: Uhlmann D.R.and Kreidl, N.J. (eds). 'Glass. Science and Technology.Volume 1. Glass-forming Systems'. Academic Press. New York, London (1983) 365–403.

Vasko, A. and Wachtl, Z.: Optical chalcogenide glasses and optical components of them. Proc. XIth Inter. Congr. Glass. Prague (1977). Vol. 5. 535–544.

Varshneya, A.K. and Petti, R.J.: Finite element analysis of stresses in glass –to–metal foil seals. J. Am. Ceram. Soc. **61** (1978) 498–503.

Varshneya, A.K.: Stresses in glass–to–metal seals. In: Tomozawa, M.and Doremus, R.H. (eds) 'Treatise on Materials Science and Technology. Vol. 22. Glass III'. Academic Press. New York, London. (1982) 242–306.

Veléz, M.H.,Tuller, H.L. and Uhlmann, D.R.: J.Non-Cryst Solids **49** (1982) 351–362.

Venghaus, H.: Integrated optics for optical communication. Glastech. Ber. **62** (1989) 175–181.

Vermylen, M.: Glass recycling in Europe. Glass Technol. **20** (1979) 80–86.

Vickers, T.A,Y.: Stronger containers require upgraded hot end handling practices. Glass Ind. **69** (1988) No.12. 15–19.

Viguié, J.C. and Spitz, J.: Chemical vapour deposition at low temperatures. J. Electrochem. Soc. **122** (1975) 585–588.

Vogel, W.: Chemistry of glasses. Transl. Kreidl, N.J. American Ceramic Society, Columbus. (1985).

Vogel, W. and Gerth, K.: Über modellsilikatgläser und ihre Konstitution. Glastech. Ber. **31** (1958) 15–28.

Vogel, W., Höland W., and Naumann, K. *et al.*: Development of machinable bioactive glass ceramics for medical uses. J. Non-Cryst Solids **80** (1986) 34–51.

Volf, M.B.: 'Technical Glasses'. Pitman. London (1961).

Wachtman Jr, J.B.: Highlights of progress in the science of fracture of ceramics and glass. J. Am. Ceram. Soc. **57** (1974) 509–519.

Wada, M.: New glass products by new forming techniques. Proc. 1st International Conference on New Glasses. December 1987. Tokyo. 65–70.

Wahl, G.: Hydrodynamic description of the CVD process. Thin solid films **40** (1977) 13–26.

Walker, A.C.: Optical computing: all-optical digital logic devices based on optically non-linear materials. Glass Technol. **28** (1987) 155–158.

Wallace, S. and Hench,L.L.: The processing and characterisation of DCCA modified gel-derived silica. In: Brinker,C.J., Clark, D.E. and Ulrich, D.R. (eds).'Better Ceramics through Chemistry' North Holland. Amsterdam (1984) 47–52.

Wallis, G.: Field assisted glass sealing. In: Doremus, R.H.(ed.) 'Glass. What's New'. Rennselaer Polytechnic Institute (1974) 176–205.

Walther, H-G. and Schäfer, B.; Herstellung und Charakterisierung von Mikroglaskugeln hoher Brechzahl mit hochwertigen Eigenschaften. Silikattechnik **37** (1986) 409–420.

Wang, S.H. and Hench, L.L.: Processing and properties of sol–gel derived 20 mol% – 80 mol% SiO_2 (20N) materials. In: Brinker, C.J., Clark, D.E.andUlrich, D.R. (eds).'Better Ceramics through Chemistry' North Holland.Amsterdam (1984) 71–77.

Wasylyk, J.S. and Southwick, R.D.: Automated statistical sampling of pressure ware. Glass (1985) no.11. 399–401.

Wasylyk, J.S.: Quality control in the glass container industry In: Boyd, D.C.andMacDowell J.F.(eds): 'Commercial Glasses' American Ceramic Society, Columbus OH (1986) 187–194.

Watanabe, M., Ono, H. and Tozyo, K.: Boosting glass container strength by surface treatment. Glass Industry **61** (1980) August 18, 23–25.

Weber, M.J.: Oxide and halide laser glasses. 2nd International Otto Schott Colloquium. Wiss. Zeitschr. der Friedrich-Schiller-Universität, Jena. Math.Nat.-Reihe **32** (1983) 239–250.

Weber, M.J.: Fluorescence and glass lasers. J. Non-Cryst. Solids **47** (1982) 117–134

Wei, Ta-Sheng: Effect of polymer coatings on the strength and fatigue properties of fused silica fibres. Adv. Ceram. Mater. **1** (1986) 237–241.

Wei, Ta-Sheng: Effect of coatings on the fatigue behaviour of optical fibres. J. Non-Cryst. Solids **102** (1988) 100–105.

Weyl, W.A.: 'Coloured Glasses'. Society of Glass Technology, Sheffield (1951).

Wicks, G.G.: Nuclear waste glasses. In: Tomozawa, M.and Doremus, R.H. 'Treatise on Materials Science and Technology. Volume 26. Glass IV'. Academic Press. Orlando, San Diego. (1985) 57–118.

Wiederhorn, S.M.: Influence of water on crack propagation in soda-lime glass. J. Am. Ceram. Soc. **50** (1967) 407–414.

Wiederhorn, S.M. and Bolz, L.H.: Stress corrosion and static fatigue of glass. J. Am. Ceram. Soc. **53** (1970) 543–548.

Wiederhorn, S.M. and Evans, A.G.: Proof testing of ceramic materials – an analytic approach for fracture prediction. Int. J. Fracture **10** (1974a) 379–392.

Wiederhorn, S.M., Evans, A.G. and Fuller, E.R.: Application of fracture mechanics to space shuttle windows. J. Am. Ceram. Soc. 57 (1974b) 319–323.

Wiederhorn, S.M. and Lawn, B.R.: Strength degradation of glass impacted with sharp particles. Part I. Annealed surfaces. J. Am. Ceram. Soc. 62 (1979) 66–70.

Wiederhorn, S.M., Marshall, D.B. and Lawn, B.R.: Strength degradation of glass impacted with sharp particles. Part II. Tempered surfaces. J. Am. Ceram. Soc. 62 (1979) 71–74.

Wilson, J.: Clinical applications of 'Bioglass R' In: Wright, A.F. and Dupuy, J. (eds) 'Glass... Current issues. NATO ASI series. Series E: Applied Sciences - No. 92'. Martinus Nijhoff. Dordrecht, Boston, Lancaster (1985) 662–675.

Wong, J. and Angell, C.A.: 'Glass structure by spectroscopy'. Marcel Dekker. New York, Basle (1976).

Worrall, A.J.: Materials for infrared optics. Infrared Physics 8 (1968) 49–58.

Wright, A.C.: Diffraction studies of glass structure. Proc. XVth Internat. Congr. on Glass, Leningrad (1989) 30–64.

Wright A.F. and Dupuy F. (eds): 'Glass-Current Issues. NATO Series. Series E: Applied Sciences. No.92'. Martinus Nijhoff. Dordrecht, Boston, Lancaster. (1985).

Wright, W.D.: 'The measurement of colour. 3rd Edn'. Hilger and Watts. London (1984).

Yamanaka, C., Yoshiaki, K. and Yasukazu, I. et al.: Nd-doped phosphate glass laser systems for laser fusion research. J. Quant. Electronics QE-17 (1981) 1639–1650.

Yamane, Y.: Monolith formation from the sol-gel process. In: Klein, L.C. (ed.) 'Sol–gel technology for thin films, fibers, preforms and speciality shapes'. Noyes Publications (1988).

Yates, J.A.: The electrical behaviour of the system for producing bronze 'Spectrafloat' glass. Glass Technol. 15 (1974) 21–27.

Young, C.G.: Glass lasers. Proc.I.E.E.E. 57 (1969) 1267–1289.

Younger, P.R.: Hermetic glass sealing by electrostatic bonding. J. Non-Cryst. Solids 38/39 (1980) 909–914.

Zachariasen, W.H.: The atomic arrangement in glass. J. Am. Chem. Soc. 54 (1932) 3841–3851.

Zallen, R.: 'The physics of amorphous solids'. Wiley-Interscience. New York, Chichester, Brisbane, Toronto, Singapore (1983).

Zarzycki, J., Prassas, M. and Phalippou, J.: Synthesis of glasses from gels: the problem of monolithic gels. J. Mater. Sci. 17 (1982) 3371–3379.

Zarzycki, J.(ed.): 'Third International Workshop on Glasses and Ceramics from Gels' J. Non-Cryst. Solids 82 (1/3) (1986).

Zijlstra, A.L.: Bibliography. In: 'Symposium sur la resistance mecanique du verre et les moyens de l'ameliorer'. USCV Charleroi (1961) 103–131, 135–204.

Zijlstra, A.L. and Burggraaf, A.J.: Fracture phenomena and strength properties of chemically and physically strengthened glass. I. General survey of strength and fracture behaviour of strengthened glass. J. Non-Cryst. Solids 1 (1968 / 69) 49–68.

Zijlstra, A.L. and Burggraaf, A.J.: Fracture phenomena and strength properties of chemically and physically strengthened glass. II. Strength and fracture behaviour of chemically strengthened glass in connection with the stress profile. J. Non-Cryst. Solids 1 (1968/69) 163–185.

Author Index

Subject Index